J. H. Tudsbery Turner, Arthur William Brightmore

The Principles of Waterworks Engineering

J. H. Tudsbery Turner, Arthur William Brightmore

The Principles of Waterworks Engineering

ISBN/EAN: 9783744649971

Printed in Europe, USA, Canada, Australia, Japan

Cover: Foto ©berggeist007 / pixelio.de

More available books at **www.hansebooks.com**

THE PRINCIPLES

OF

WATERWORKS ENGINEERING

BY

J. H. TUDSBERY TURNER, B.Sc.

HUNTER MEDALLIST OF GLASGOW UNIVERSITY
MEMBER OF THE INSTITUTION OF CIVIL ENGINEERS

AND

A. W. BRIGHTMORE, M.Sc.

LATE BERKELEY FELLOW OF THE OWENS COLLEGE, MANCHESTER
ASSOCIATE MEMBER OF THE INSTITUTION OF CIVIL ENGINEERS

London:

E. & F. N. SPON, 125 STRAND

New York:

SPON & CHAMBERLAIN, 12 CORTLANDT STREET

1893

PREFACE.

THE art of the Civil Engineer consists, to use the
language of Thomas Tredgold, of the " practical application of the most important principles of natural
philosophy."

Whilst examples of sound engineering practice
are within reach of all, the laws that govern it
are frequently obscure. An orderly statement of
the facts ascertained by experience, will, however,
throw light upon the principles involved. It has
been our aim to investigate and classify the observations and rules employed, and to trace the steps
taken by the engineer in proceeding from the inception to the completion of schemes of water-supply.

From the data afforded by experience of executed
works, and of the properties of materials, designs
appropriate to the requirements of varied circumstances may be elaborated ; but the fixed laws of
natural philosophy must not be ignored by the
designer, in applying the conclusions derived from
his observation of facts and phenomena.

These considerations have led us to neglect the historical aspect of the subject ; because during many ages, limited mechanical and metallurgical knowledge and an imperfect acquaintance with the laws of hydraulics, prevented the ancients from leaving to us many examples of waterworks which can be prudently copied.   It is only by scientific investigation of the circumstances of individual cases, that false analogies are avoided between more recent practice and the requirements of works contemplated. " The site of a spring or the presence of a stream," says Professor Prestwich, " determined probably the first settlements of savage man ; and his civilised descendants have continued until the last few years equally dependent upon like conditions."   The resources of nature are by the skill of the engineer now rendered more accessible to mankind ; and communities are enabled to obtain water without regard to the depth and remoteness of its source.

In the following pages some knowledge of physical laws is assumed, and we endeavour by their aid to set forth the principles of the design of waterworks —illustrating these by examples of recent practice. A comprehensive grasp of the subject will be best afforded by the consideration of its separate elements. Each one of the eight chapters in which we seek to develope an outline of this branch of engineering, comprises a statement of matters which, if not necessarily related to one another, are conveniently grouped together—the several articles being rendered as far as possible independent, but referenced one to another where their subjects naturally touch.

The first chapter treats of the " sources of water-supply," and of the preliminary considerations involved in the selection of such sources. In the second chapter the " measurement " of water is introduced, forming the foundation of what may be termed the science of waterworks engineering. " Collection," in the third chapter, treats of the methods employed in obtaining access to sources of supply, with an account of the appliances used to abstract water therefrom. The fourth, fifth and sixth chapters comprise respectively the " storage," " purification," and " conveyance " of water—the last subject embracing the design, construction and use of aqueducts of various kinds. The important questions concerning the " distribution " of water, occupy the seventh chapter ; and the book is concluded by a brief statement of some important considerations that affect the " maintenance of waterworks."

For many of the mathematical methods and expressions employed, we gratefully acknowledge our debt to previous writers. Many are original and appear here for the first time. In particular, the subject of masonry dams is treated by a new method—the object of which is to determine by easy

calculations an economical dam-section that rigidly
conforms to the primary conditions of strength and
stability laid down at the outset of the investigation.
The extent to which mathematical treatment is
desirable in a work of this character must be largely
a matter of opinion ; but there can be little doubt
about its utility in often presenting more concise and
explicit statements than are afforded by verbal
expression. To this extent we have employed it—
endeavouring to take advantage of its aid, but not
to magnify its importance.

We have not sought to describe operations that
are common to every kind of engineering construction ; and in occasionally adverting to details that
are peculiar to this class of undertakings, we would
disclaim any attempt to teach arts that can only be
acquired by contact with the work itself.

Our object has been to provide, for the use of
those engaged in the construction of waterworks, an
orderly and logical statement of the matters that
specially occupy the attention of the engineer ; presenting the subject in such a form as to be both
serviceable to students, and instructive to that large
class of persons who desire some general but accurate knowledge of a subject to which their attention
is compelled by the sanitary progress of this age.
In alluding to executed works, elaborate descriptions
have been avoided ; their proper place being, in our

opinion, in the pages of the scientific and professional journals, where, for the most part, they originally appear.

Every acknowledgment is tendered for the great assistance we have obtained from the numerous publications referred to in the text—particularly those of the Institution of Civil Engineers. The wide range covered by the practice of the hydraulic engineer, and a desire to state within a reasonable compass the principles that guide him, may, we trust, afford an apology for such want of elaboration and detail as the reader may feel in the perusal of these pages.

J. H. T. T.
A. W. B.

London : *June*, 1893.

CONTENTS.

CHAPTER I.

THE SOURCES OF WATER-SUPPLY.

CHAPTER II.

THE MEASUREMENT OF WATER.

CHAPTER III.

THE COLLECTION OF WATER.

THE PRINCIPLES

OF

WATERWORKS ENGINEERING

———◆◆◆———

CHAPTER I.

THE SOURCES OF WATER-SUPPLY.

1. IN approaching the subject of waterworks engineering it will be useful first to consider several important
properties of water, the conditions that result from these,
and the influences they exert upon other kinds of matter.

The most valuable attribute of water, from an engineer's
point of view, is the fluidity which enables it to be readily
conveyed from place to place by its own gravitating impulse. This, however, is a property by no means peculiar
to liquids, but possessed in some degree by all matter—
a fact that presents itself in the course of every application of engineering art. In the construction of works designed to support severe and prolonged stress, phenomena
such as that exhibited by the plasticity or ductility, as
it is termed, of iron and steel,* may well afford a warning
against the distortion to be apprehended from the action
of unsupported stresses upon the molecular constitution of
materials that are popularly regarded as rigid.

* *Vide* Tresca on ' Flow-of Solids,' Proceedings Inst. Mech. Eng., 1867,
p. 114 ; 1878, p. 301.

B

2. Among other properties of water may be noticed :—
(1) Its high degree of incompressibility; whence proceeds the phenomenon of its density remaining constant
under the pressures to which it is subjected in ordinary
practice. To this is due both the readiness with which
solid matter that will not float upon its surface is precipitated through it when at rest, and the formidable character
of its impulsive force when in motion. (2) Its viscosity,
which is sufficient to render possible its confinement by
structures composed of distinctly permeable materials,
such as stone and earth ; and by virtue of which the great
natural sources of water supply existing beneath the earth's
surface are retained and rendered accessible to man. (3)
Its great capacity for heat and low conducting power—the
joint causes of that remarkable equability of temperature
evinced by it under alternations of atmospheric heat and
cold. (4) Its low expansibility ; whence its volume alters
but slightly whilst passing through a considerable range of
temperature. This property enables it to be confined in
closed vessels and other apparatus without sensible variation of pressure at ordinary atmospheric temperatures.
(5) Its characteristic power of dissolving all gases and
almost all other substances at certain temperatures, and
of expelling them from solution at other temperatures.

3. All these properties affect the question of the design
and construction of waterworks, whether for purposes of
irrigation, power, canal, trade, or domestic supply. But to
the last-mentioned action must be ascribed the origin of
some of the gravest difficulties encountered in procuring
sources capable of yielding supplies of pure water. Although,
if scientifically developed, the subject would involve considerations unsuited to the character of this work, we will
here summarise the main points relating to the constitution
of water, so far as they affect the practice of the engineer.

4. Water, although an indifferent oxide, is remarkable inasmuch as its combining tendencies extend over a wider range than those of any other compound. Its properties as a solvent have been already alluded to. Hence, from the instant of its generation in the form of mist or rain, it is susceptible to pollution ; from which it can only be completely freed by an evaporative process more refined than that which distils, in nature, the water of the clouds from the saline mixtures of the ocean. As mist, it dissolves the gases that constitute the atmosphere ; and, in the neighbourhood of large towns, these are important elements of impurity. In falling to the earth as rain, it further carries down in suspension the dust and soot that are present in the air. Upon reaching the earth, it encounters the risks of animal, vegetable and mineral pollution in various degrees, depending upon local circumstances. The kind and amount of such pollution affect both the value of water for household and manufacturing purposes, and its attitude towards the works designed to store or to convey it.

5. In seeking supplies of water, the natural and customary practice is to pursue the search in those directions that present the least likelihood of contamination from either natural causes or industrial operations.

The successful prosecution of such a quest requires the application of scientific knowledge of a wide range, alertness, and experience in the work. It is necessary to embrace in one glance the physical features and geological structure of the district under review, as well as the class of agriculture and its extent, the density of the population and the sanitary conditions that obtain among them. There are, moreover, many polluting agencies whose existence is not immediately obvious, and the influence of which cannot be determined by superficial examination ; consequently, a just opinion of the value of any given water, and especially the propriety of using it for domestic supply, must be formed not only upon evidence afforded by inspection of its source, but also upon that yielded by enquiry into its composition.

Such enquiry may be chemical, or microscopical, or bacteriological, or a combination of any of these methods. The important matter is that the results of the examination shall be interpreted aright. It is a common practice to judge the relative merits of water upon the basis of chemical analyses or bacteriological statistics ; but, when such particulars are presented for the consideration of persons ignorant of the methods by which they have been arrived at, and when no standard relation can be assigned between the various constituents revealed by the chemist and the biologist, prudence will suggest to those unacquainted with the intricacies of the subject the propriety of reserving the formation of opinion upon such obscure evidence.

6. The chemical analysis of water reveals, indeed, the elements that compose it, but fails generally to indicate precisely the state in which those elements are combined together. The presence of certain metals, such as lead and copper, in any amount exceeding 1 part in 1,000,000 parts of water, is sufficient to stamp it as unfit for domestic use, without the need for further enquiry as to what particular compounds of those metals are present. The "hardness" of water is regarded as an impurity only if it exceeds certain limits, and, except in connection with the process of softening (§ 306), the question is hardly raised as to whether " hardness " is due to the presence of the carbonate or the sulphate of lime or some other alkaline or metallic salt.

But when we pass from simple mineral impurities, to a consideration of the organic pollution of water as indicated by the presence of inorganic substances in it, the problem becomes one of extreme complexity—we even venture to say that it is indeterminate. In stating this, we would not for a moment suggest that the chemical examination of water is not of high value. On the contrary, it is a necessary proceeding, furnishing one portion of the evidence that is indispensable to the formation of a sound judgment on the quality of any given source of supply.

7. All waters contain dissolved oxygen and carbonic acid derived from the atmosphere ; but these substances, although foreign to pure water, are far from being regarded as " impurities," in the sense to be attached to that term in reference to water-supply.   Ammonia and the acids of nitrogen, chlorine and sulphur are also, under some circumstances, derived from the atmosphere ; but, when any of these are present in appreciable quantity, they distinctly constitute impurities. The reason for this is rather that they afford evidence of, and tend to promote, organic pollution, than that they can be considered, in the minute quantities in which they occur in nature, as themselves directly injurious to animal life. Other matters generally present in water are the soluble mineral impregnations of the strata over or through which it has flowed after falling upon the earth. Of these, the salts of the alkalis, alkaline earths and iron are not regarded as injurious, in the amounts in which they are ordinarily found, or to which they may be readily reduced by some simple treatment— although their presence may be a source of inconvenience and involve expense to the consumers. When such salts occur in extraordinary quantities, as for example, in the Triassic marls of East Cheshire, where the ground-water is frequently far too much impregnated with sodium chloride for ordinary use, they may lead to the absolute rejection of certain sources of supply. In other familiar

instances, excessive mineral impregnation, whilst causing water to be unfit for ordinary domestic use, renders it peculiarly applicable to medicinal purposes. The rest of the soluble metallic salts cause unqualified pollution, but are fortunately seldom present in ordinary water.

The dangerous forms of impurity against which it is ordinarily necessary to guard, are in reality the organised growths that abound in water tainted with decomposing vegetable or animal matter. The *algæ*, which in the light organise their own food from the nitrogenous compounds and the carbonic acid dissolved in water, furnish means of subsistence to the animal life in water as well as to *fungi* and to certain forms of bacteria, which lack the power of food-organisation possessed by the *algæ*. The determination of the nitrogen present in its several forms is then of the first importance. Next comes the estimation of that portion of the chlorine present which may be regarded as the result of animal pollution. All natural waters contain chlorine—as a constituent of common salt, it is a daily article of food—but, as it is also a constituent of animal excreta, its presence in abnormal quantity in a given water may well give rise to a suspicion of sewage pollution. By "abnormal quantity" is meant, of course, not necessarily a large quantity, but an amount in excess of that found in water known to be good, and derived from the source or district under consideration. Without such information as this, a statement of the chlorine present is valueless as evidence of the quality of water.

The nitrogenous elements of water are now generally stated as fractional weights of ammonia, albumenoid ammonia, nitrites, and nitrates. Ammonia exists usually as chloride or carbonate, and, in good water, seldom exceeds the minute quantity of 1 part in 20,000,000. "Albumenoid" ammonia indicates the undecomposed nitrogenous matter in solution, which should not exceed 1 part in 10,000,000. The statement of the nitrites and nitrates affords an indication of the extent to which mineralisation of the nitrogen originally present has proceeded under the action of the micro-organisms contained in the water. The ratio borne by the organic carbon to the nitrogen is sometimes given, as suggestive of the probable vegetable or animal origin of the nitrogen —the latter kind being generally deemed the more dangerous to health. It may, however, be observed that, according to such a system, the mixture of two waters containing very unequal ratios of carbon to nitrogen would lead to an apparently unintelligible result.

8. Other qualities of water, which, although not ascertained by chemical analysis, are frequently determined at the same time, are the colour, odour, total amount of solid matter present, and the "hardness" expressed according to Dr. Thomas Clark's scale.*

9. In taking samples for the examination of water, stoppered glass bottles should be used ; and both bottles and stoppers should be first well rinsed with the water in question. The temperature of the water and of the atmosphere should be observed and noted. After being filled, the bottles should be at once sealed and numbered on the seal—a note being made to identify the number, stating the date, time and place, and the name of the collector of the samples. The selection of the times and places at which samples of water should be collected, exercises, especially in the case of rivers and streams, a most important influence upon the conclusions to be derived from their subsequent examination. The object aimed

---

* The degrees of Clark's scale correspond with the number of grains of calcium carbonate, or other equivalent soap-destroying salts, in one gallon of water.

at is to arrive at the characteristic quality of the water at the source of supply. Agricultural and other industrial operations, the growth and decay of plants, the surface denudation wrought by floods, and the dilution resulting from continuous rains, are a few of the ever-varying elements that affect the quality of water. In order, therefore, that analysis or other examination may be of practical value, series of samples must be collected over a considerable period, and the information derived from these must be interpreted with due regard to the several conditions under which they were taken. General conclusions drawn from the results of single analyses made at a particular time, and based upon an unwarrantable assumption of the constancy of composition of the water at all times, are almost valueless, and may be directly misleading.

10. Thus far, we have dealt mainly with the chemical examination of water, which shows : (1) the minerals present—of whose hurtfulness or otherwise ample knowledge exists ; and (2) the presence of those organically-derived elements which are known to be, under favourable circumstances, essential to the development and subsistence of organised growths in the water. The question that naturally follows is, Does the water in question contain growths susceptible to such development, and, if so, are they of a character injurious to health ? The former part of the question is readily answered by the microscopist ; who is able, moreover, to give a satisfactory account of the properties of the vegetable substances in the water disclosed by his enquiry. In general, it may be stated that the less these are, the better.

11. With regard to the bacteriological examination of water, it must be admitted that it can hardly yet be relied upon to furnish exact and entirely satisfactory results. Certain important facts are, however, well established.

All natural waters, unless sterilised by some appropriate process, may be said to contain micro-organisms that subsist upon dissolved or suspended substances ; and the quality of a water exercises a potent influence upon the growth and predominance of particular species among those that inhabit it. The discovery of pathogenic forms, such as those of the *spirillum* of Asiatic cholera and the typhoid *bacillus*, among water-microbes, has, whilst confirming belief in the propagation of zymotic diseases by their agency, led to investigations into the conditions of the vitality of many well-known pathogenic organisms, in waters of different composition under various conditions. The main object of bacteriological examination, as at present practised, is to ascertain the number and the varieties of well-known forms of microbes contained in the water ; and is effected by the process of gelatine-culture devised by Dr. Koch, of Berlin. But the number of colonies of bacteria present in a given volume of water cannot be regarded as a satisfactory guide to its wholesomeness ; and the presence of a large number of species can merely suggest the probability of some previous contamination.

The real test of the value of the water, in this connection, must be sought in the specific determination of the species present in it, and the tendency of their environment to promote increase or decrease of their numbers. The complexity of the problem is such that the results of any such examination could only be usefully interpreted by a person skilled in biological science ; and, among the many elements that must be considered, the following may be cited to illustrate this point.* Water affects the living protoplasm mechanically, physically and chemically ; and all the substances dissolved and suspended in it exert

* First Report of the Water Research Committee of the Royal Society, by Drs. Frankland and Marshall Ward. Proc. Roy. Soc., vol. li. p. 215.

pronounced actions on the organism.   The temperature of the water is of the utmost importance as regards the life of any given species; and light is a factor that must be considered in many cases.   In addition to this, it is highly probable that every ripple on the surface of lake or stream, every pause in a back-water, and every fall over a weir or cascade, affect the matter, if only in altering the gaseous contents of the water, or the distances between the individual micro-organisms.

12. From these considerations, it is plain that the arguments in favour of relying only upon long-continued series of analyses, to test chemically the character of any given source of supply, apply with increased force when the biological character of the water is in question.   And, indeed, having regard to the fact that the most dangerous kind of organic impurity may occur fortuitously in waters of the highest class, it may be safely concluded that the function of the biological examination of water does not by any means end with the preliminary investigations that are undertaken in the selection of sources of supply.

13. No attempt is made to consider the question of a general standard of purity, or to do more than indicate the lines upon which water-examination is, or should be, conducted.   Amidst the difficulties that beset analytical enquiry into the purity of water-supplies, it is satisfactory to find that, in practice, chemical and microscopical investigations seldom afford grounds for condemning water derived from sources that have stood satisfactorily the test of a critical inspection.   To aid and prompt the engineer in his enquiry, however, these sciences occupy a trusted position ; and must be appealed to in every case where the sanitary properties of water are under review.   Further, in those cases in which the application of processes of purification is required, as will appear in a later section of

this work, the design and efficient working of the methods employed depend largely upon an accurate knowledge of the amount and kind of the impurities under treatment. In order, therefore, to form an opinion as to the necessity for subjecting water to a process of purification, as well as to determine its nature and to estimate the cost of its application, it is essential that the engineer be furnished with proper information upon the constitution of the water that he proposes to supply, whether it be for domestic or industrial use.

14. The several sources of supply known to hydraulic engineering science are to be regarded as merely stages of the various courses pursued by water in its passage from the rain-clouds to the ocean. Whether precipitated through the atmosphere as rain, or flowing over the earth's surface as stream or river, or percolating the soil and rocks beneath, the motion of water is to be explained according to the same uniform physical laws.

15. At the risk of being tedious, this proposition must be rigidly demonstrated ; and the unseen motion of water in its continual circulation must be shown to be only a variation of its visible mode of flow in streams upon the earth's surface. The use of the "divining rod" is still resorted to,* and many persons unconsciously attribute to subterranean waters a certain independence of ordinary physical law. Few poetic images surpass in beauty that presented by the descent of those rivers of antiquity from the mountain to the sea, obedient to nature's law :—

> " At sun-rise they leap from their cradles steep
>   In the cave of the shelving hill ;
> At noon-tide they flow through the woods below
>   And the meadows of asphodel ;

---

* Whilst this article is passing through the press, we read that the services of a magician have been employed to find water in certain parts of Lancashire during the recent drought, June 1893.

And at night they sleep in the rocking deep
    Beneath the Ortygian shore ;
Like spirits that lie in the azure sky,
    When they love but live no more."

But the unnatural process to which was ascribed in the
same poem the ascent of the water to the mountain-tops,
requires for its acceptance mythological lore or the credulity
that springs from ignorance. The lurking tendency to a
belief in irrational modes of the occurrence and motion of
water underground, which is responsible at times for the
waste of large sums of money, is not easily dispelled.

16. It is a common practice in tropical and semi-
tropical regions to catch rain as it falls on the roofs or
floors of vast tanks ; and in many places it is customary
to lead rain-water from the roof of buildings and to store
it in cisterns. Thus obtained, water is charged with atmospheric impurities only. It may answer in rural districts
and for irrigation purposes ; but, in the neighbourhood of
large towns and in manufacturing districts, such sources of
supply for domestic use are open to the gravest objection.

17. After falling upon the surface of land in the form
of rain, water, subject to some losses that will be alluded
to hereafter, still continues, under the action of gravity,
to seek lower levels, pursuing those routes in which it
experiences the least resistance to its downward motion.
Generally speaking, this direction is vertical through soil
and sub-soil, until its progress is checked by encountering
the great body of water that saturates the subterranean
regions at depths depending upon local circumstances.
Here the motion is not arrested, but its direction becomes
inclined at a certain angle, which is determined by the
resistance opposed to the flow by the strata at the place
in question. The inclination of this subterranean water-
. slope changes from point to point according to the geological formations traversed. In permeable rocks, such

as chalk* and gravels, the slope is naturally flat ; in sandstone it is less so ; whilst in compact grits the angle of
inclination is large—in just such degree as the impervious
character of the rocks requires greater hydrostatic force
to overcome their resistance to the passage of the water.
The absolute level at any point of the slope also variès
according to the volume of water which, contributed by
the rainfall, seeks a passage to the ocean.

Thus there extends in all directions a "surface of
saturation," or, as it is sometimes termed, "plane of saturation," occurring always where the descending waters assume
a definite surface-slope, in accordance with ordinary hydro-
dynamical laws. On the whole, this surface of saturation is inclined towards the sea, although circumstances
frequently produce local inclination in other directions, so
marked as to mask the general character of the slope.

The surface of saturation ordinarily coincides with
the "free-level" of the subterranean waters at every point
in the district, although in synclinal basins overlaid by
extremely impervious formations this is not necessarily
the case. In a district, the geological structure of which
is of a compact and impervious nature, the surface of
saturation is often situated at no great depth underground, and may at times, when the rainfall is heavy,
become raised until it coincides with the land-surface—
such a condition resulting evidently from the permeability
of the land being barely adequate to meet the demands
of such increased quantity of water for a passage through
it. The land is then said to be "water-logged." There
are in general two conditions of which the immediate
result is the establishment of flow upon the surface.

18. Case I.—When the surface-slope of a considerable

* The permeability of chalk is principally due to the fissures that traverse it.
(§ 131).

tract of land is less than the hydraulic gradient* required
to force the entire volume of water through the earth
as rapidly as it falls upon it, the surface of saturation of
the district rises above the surface of the land, until a
hydraulic gradient is formed adapted to the circumstances
of the case, and part of the flow takes place over the
ground. The "hydraulic surface," as the free surface thus
formed may be conveniently designated, does not differ
much from the ground-surface, because the water flowing
above ground is comparatively free from frictional resistance, and a slight fall is enough to produce considerable
velocity, and to effect discharge off the surface as fast
as the rain falls upon it.

In the special cases of rain falling upon frozen ground,
or falling very heavily, the resistance of the surface to
the passage of water through it may be so high as to
prevent any considerable portion of it from penetrating
the earth; and abnormal flow may be established upon
the surface, although the true surface of saturation is at
the time situated at some depth beneath. Those who have
experience of severe tropical rains, or of floods caused by
the sudden melting of large accumulations of snow, must
have been astonished to observe the current and the depth
of water that may temporarily prevail over wide areas
of land into which ordinary rainfall disappears at once.

When the surface of saturation is high, the smallest
depression in the land may be sufficient to cause it to
issue therefrom, since the water-slope in any direction is
determined by the facilities afforded to the passage of
water in that direction. Any hollow below the surface

* The hydraulic gradient, inclination or declivity is a conception explained
exactly in treatises on mechanics. It is a line tangential to the "free-level"
surface of water in motion, and measures by its curvature at any place the
variation of the resistance to flow there experienced by the water.

of saturation presents to the water in the adjacent ground a course of diminished resistance which is naturally taken advantage of. In proportion to reduced resistance the surface of saturation becomes flattened, until, in the hollow, it issues above the ground as a true hydraulic surface.

The rills on every hill-side, no less than rivers and lakes, owe their origin and maintenance to such causes ; and it is due jointly to the high position of the surface of saturation and to their undulating character, that districts of hard and impervious geological structure lend themselves so readily to yield "surface-water"—that is to say, water which, after falling upon the earth, is almost at once directed, by its own gravitating impulse, to flow in channels on the surface of the land. Such waters, if derived from hard rocks in a sparsely-populated and agriculturally barren district, are generally the least open to objection of those yielded by any sources of domestic supply, and are also well adapted for irrigation purposes. In quality, they enjoy comparative immunity from animal and vegetable pollution ; the quantity of water available can be determined with a high degree of accuracy, and the water-rights, or even absolute proprietorship, of the catchment-area or watershed* may generally be secured at a moderate cost. The two last-mentioned attributes are of great importance in schemes of water-supply for power-purposes.

From the circumstances considered, it is evident that the surface-water flowing off any land by no means comprises all the water that gravitates away from the area in question. No formation is absolutely impervious, and all rocks permit a certain quantity of water to pass through their interstices towards the sea at rates depending upon

* The term "watershed" is sometimes applied to the "water-parting" or boundary of a catchment-area.

their permeability. The quantity thus transmitted invisibly beneath the earth's surface is termed, with respect to the water of any catchment-area, the "loss by percolation" (§ 60), and is the origin of that collected in wells.

In tracing the further course of water from its first contact with the earth, it might seem appropriate here to allude to the two natural processes of "evaporation" and "absorption" by which water is lost—i. e. rendered unavailable for supply-purposes. These losses are, however, duly dealt with under "measurement of water"; and it will be sufficient at this point to direct attention to the fact that evaporation is a process that affects, of course in varying degrees, surface-waters, subsoil-waters, and rivers; and that "absorption" or the combination of water with either the minerals of the earth or with the other elements of vegetable and animal tissue, affects in the former case chiefly sub-soil, and in the latter, surface waters —resulting, to a certain extent, in their gradual withdrawal from circulation.

Loss by percolation may be permanent or merely temporary. In the former case, the configuration of the land, conditions of rainfall, and resisting power of the sub-soil to flow through it, have a joint result in maintaining the surface of saturation below the ground. In the latter case, some portion of the flow issues from the land at lower levels, and joins the current of the true surface-water. Hence results the second condition under which flow is established upon the earth's surface—already indicated in the description of the formation of the rills on a catchment-area.

19. Case II.—When, at any place, the surface-slope of the land is of higher inclination than the hydraulic gradient required by the flow of the percolating sub-soil waters through the rocks, the surface of saturation naturally issues

above the ground in the manner described in § 18. Illustrations of this action frequently occur in the streaming vertical faces of sandstone quarries, and in the marshy areas often found on steep hill-sides. Some of the water that enters every ditch is contributed in like manner from the adjacent sub-soil.

20. There is another characteristic mode of the formation of springs, which is a composite result of the actions considered in Cases I. and II. If, on a sloping land-surface of permeable character, we pass by faulting or otherwise, on to the outcrop of a comparatively impervious rock barrier, *Fig. 1*, the surface of saturation *s s* in the permeable formation A, will be raised to furnish the steep hydraulic gradient required in the more compact rock B, beyond ; and may be so far elevated as to issue

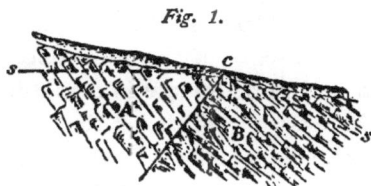

Fig. 1.

from the ground at the line *c* of the barrier—part of the sub-soil flow emerging there as a spring, overflowing the lip of the permeable basin formed by the barrier. The application of the principle that the surface of saturation must be everywhere determined by the frictional resistance of the rocks and the quantity of water that seeks a passage through them, is sufficient to explain fully all questions connected with the action of springs that arise in practice.

Intermittent springs, "bournes" and "gypsey-races," result from temporary abnormal rise of the surface of saturation, consequent upon heavy rainfall, and its issue above the ground. The disappearance of streams, whether gradually, owing to the slope of the beds being flatter than the hydraulic surface, causing the re-formation of the surface of saturation underground, *Fig. 2*, or by "sinks"

C

and "swallow-holes," *Fig. 3,* the occurrence of which causes depression of the hydraulic gradient so suddenly as to form steps in it, may be accounted for on the same principle; and could be accurately predicted if we possessed a proper measure of the retarding influences exercised by the strata under consideration, and of the quantity of water to be dealt with.

Fig. 2.

Fig. 3.

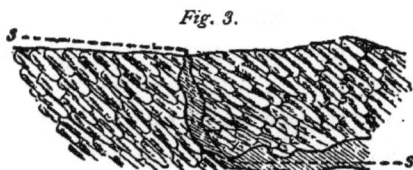

21. Although the subject of underground waters is, from absence of the means for exact measurement, necessarily indeterminate in a mathematical sense, much of the obscurity in which, at the outset, the important practical problems connected with springs as sources of water-supply appears to be involved, may be removed by the application of simple physical laws—interpreted, of course, with due regard to the geological structure and the meteorological conditions of the district under review. The yield of springs naturally varies with the inclination of the hydraulic gradient or the level of the surface of saturation to which their occurrence is due, and is entirely dependent upon the rainfall of their contributing areas. It is a quantity capable of estimation with some precision, if all the data are properly ascertained. Nothing can be more unsafe than to rely upon statements unsupported by recorded observations, or upon local tradition, when the capabilities of a spring are in question. Even springs that are truly perennial, are invariably intermittent as regards the quantity of water discharged by them at different seasons; and caution must always be observed

to avoid confusing permanence of action with constancy of supply.

Spring-waters are always impregnated with the mineral solutions of the rocks from which they issue ; and, if these impregnations are not excessive, the water is generally, owing to the absence of organic contamination, most suitable for domestic use. A catchment-area, the geological structure of which lends itself to the production of springs, is valuable both on account of the purity of its dry-weather flow, and of the natural storage which exists within its rocks—maintaining a gradually diminishing, but nevertheless continuous, yield of water during rainless periods, until the surface of saturation falls to the level of the lowest spring. The water is generally hard, and frequently contains a somewhat excessive quantity of iron —2 parts in 1,000,000 of this metal should be regarded as a limiting amount ; but these are impurities amenable to simple treatment (§ 268), and can only be said to constitute pollution when present in unusually large quantities.

22. The next stage in which water is found, as a source of supply, is flowing in rivers—channels that maintain a perennial though ever-varying discharge. The formation of a river is due to precisely the same cause as that of the smallest rill. It owes its maintenance to the rainfall of its district preserving the level of the surface of saturation above the natural hollow that forms its bed. The occurrence of river-valleys, small originally, but ever widening and deepening by the erosion due to the scour and fretting of their currents, offers to the water percolating through adjacent land, a course of less resistance than that of the interior of the rocks ; the subterranean waters gravitate towards the bottom of the valley ; the surface of saturation is depressed in the vicinity, rapidly at first but flattening as the river is approached, *Fig. 4,* and emerges

from the ground coincident with the hydraulic surface of the river.   The water flowing in rivers is contributed in

Fig. 4.

three ways; directly, by adventitious surface-flow, and by rain; indirectly, by rivulets and ditches, which tributaries derive their own flow as miniature rivers; and, normally, by the percolating land-water that enters their beds under the hydraulic head of the neighbouring subterranean waters.   The last-mentioned form of contribution is sometimes peculiarly marked by the evident increase in the size of rivers, without the apparent cause that is afforded by the junction of tributaries.   Thus, in defining the watershed or catchment-area of a river, it is necessary to consider not only the superficial extent of land that discharges surface-water into it, but, further, the area from which underground water is contributed to it—two elements that are seldom co-incident.

The quality of river-water is seldom unexceptionable, and is frequently bad.   Flowing through land the fertility of which is identified with their existence, receiving pollution at the hands of all for whom they form a natural resort,* and carrying in suspension the detritus of their several districts, rivers are generally charged with impurities of vegetable, animal, and mineral origin, to such an extent as to render them frequently a most desirable

* Prof. Joseph Prestwich, Presidential Address to the Geological Society, 1872.   Quarterly Journal Geol. Soc., vol. xxviii.

source of water for irrigation purposes, but not for domestic supply. Still, by certain processes, even the most unpromising river-water is often rendered generally serviceable.

23. The self-purification of streams during their flow has engaged much attention ; and, although it must be conceded that such action does take place, it is infinitely less effective than the natural processes of filtration and distillation. In dealing with statistics bearing upon this subject, apparent purification due to mere dilution must be carefully distinguished from real abstraction or alteration of deleterious elements. One of the principal conclusions arrived at by the Royal Commission of 1868, appointed to consider the best means of preventing the pollution of rivers, was : "There is no river in the United Kingdom long enough to effect the destruction by oxidation of sewage put into it at its source." The exhaustive nature of the enquiry that formed the basis of that conclusion entitles it to careful consideration ; and, although during the last twenty years much additional knowledge of the operations that take place in rivers has been gained, it does not lead us to believe that aëration and the agency of plant and animal life produce a much more marked effect than that of the limited extent formerly assigned to them. The result of the aëration produced by even the Falls of Niagara is so insignificant as to be insensible so far as regards any purifying effect upon the water.*

24. Somewhat closely allied with river-water, but superior to it in many respects, is that derived from shallow wells. Where it happens that the surface of saturation is not very far below ground, as is always the · case with flat and low-lying land near the sea or rivers, and sometimes on extensive plains far distant from the

* Journal American Chem. Soc., Nov. 1890.

sea, the waters may be intercepted in their onward course by shallow wells. The yield of such wells is governed by the rate of percolation through the land ; and water may be drawn from them quite uncontaminated by the impurities of the neighbouring river- or sea-water, provided that the draught from the well does not exceed the natural rate of percolation through the ground in the normal direction of flow. The "filter-galleries" frequently used in the United States and elsewhere, only act partially as intercepting wells—their supply being drawn by heavy pumping mainly from the adjacent rivers. In the neighbourhood of tidal basins, the water-level of such wells is observed to oscillate periodically with the tide, even though they are situated above the sea-level. This effect is due to periodical abnormal elevation of the surface of saturation, and flattening of its slope by the checking or impounding of the subterranean flow during the rise of the tide. During its ebb, the slope becomes steeper, and the rate of percolation is increased and the surface of saturation lowered. The land-water is thus discharged by pulsations corresponding with the tidal period. Any attempt, however, to pump water from such sources more rapidly than it is contributed to them by natural percolation through the earth in the normal direction of flow, has the inevitable effect (intentionally produced in the case of "filter-galleries") of reversing the flow from the natural outfall, which then contributes water directly to those sources —in a more or less perfectly strained condition, according to the compactness of the strata in which they are situated.

The water yielded by shallow wells is generally clear, and in wet seasons is fairly good. The percolation of the water through the capillary pores of the earth deprives it to a great extent of the living matter with which surface-waters are peculiarly abundant. The absence of light, aided by other agencies, such as temperature and the nature of the mineral solutions present in the underground water, promote the sterilising process ; which, however, is not so complete as that effected in spring-water and in that derived from deep wells, to be subsequently noticed. The principal danger to be apprehended in the class of water now under consideration, is that arising from fouling by surface-drainage and animal excreta, which, particularly in dry seasons, sink into the ground and directly join the flow of the underground water. Hence the necessity for lining all such wells down to and somewhat below the lowest level of the surface of saturation in their locality.

25. Deep wells may be described as those which, passing far below the surface of saturation, draw their supply from strata that are presumably entirely unaffected by surface impurities. Where such wells pass through permeable strata from the surface downwards, they are steined or lined to exclude surface-water from their immediate vicinity. They are frequently employed to reach permeable water-bearing strata underlying, and possibly also underlaid by, dense and impervious formations ; and where those strata form the base of a synclinal basin, the waters with which they are charged rise under the hydrostatic pressure exerted in the more elevated parts of the basin, and may completely fill the well or bore-hole, and overflow at its surface. When this takes place, the wells— then termed Artesian—afford the most satisfactory result of the well-sinker's art. It has been estimated by Professor De Rance that there are 26,633 square miles of porous rocks in England and Wales, and in addition, 19,308 square miles of impervious clay overlying permeable rocks. These figures indicate the importance that must necessarily attach to the subject of supply from the underground waters of this country, which, unlike surface-waters, do not involve

great storage-reservoirs and the consequent withdrawal of considerable areas of fertile land from agricultural use.

26. The choice of the proper situations for such borings demands an exact knowledge of the geological features of the entire district, and a full comprehension of the mineralogical and stratigraphical conditions that affect the question of the abnormal subterranean flow set up by opening the wells. Misconception of the phenomena concerned sometimes leads to the commission of costly errors in this respect. A notable example was some years ago furnished by the large bore-hole at Bootle, near Liverpool, which at great expense was sunk to a depth of 1300 feet in response to a certain popular demand, with the main result of proving the absence of any abnormal occurrence of subterranean water in the Triassic strata of that district—a fact that was at the time urged upon the attention of the townspeople by their professional advisers, and should have required no such demonstration.

27. Where a deep well does not pass through any impervious band, but draws its supply from the superficial rocks of the earth's crust, pumping produces a local depression of the surface of saturation, which extends in

Fig. 5.

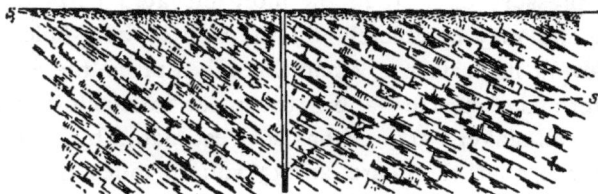

every direction round the well, forming an inverted " cone of depression," or " cone of abstraction," *Fig.* 5—convenient but very inaccurate terms for the figure assumed by the surface of saturation around the well. After a deep well has been sunk, the formation of the "cone of depression" proceeds *pari passu* with the rate of pumping practised. By degrees the apex of the cone approaches the bottom of the well; and, in order to maintain the yield, its depth must be increased. A limit, however, is presently reached, for the yield of the well, becoming gradually more dependent upon the contribution of rainfall derived from the base of the "cone of depression," begins to feel the effect of contamination from the surface-waters that now fall as it were directly to the apex of the cone, and are drawn into the well by the slightest excess of pumping-effect over the supplying capacity of the normal subterranean flow now barely touched by it.

Such is the history of nearly all wells situated in the neighbourhood of large towns, which have generally to be at last disused, owing to contamination of their waters by surface-impurities; although their life may be greatly prolonged by refraining from excessive pumping at any time—by which means the supply derived from them may be, over a long period, in the aggregate largely increased. A noteworthy example was afforded some time back by the Dudlow Lane well, belonging to the Liverpool Corporation. This well, sunk to a depth of 247 feet in red sandstone, and, including tunnels or galleries, having an area of 2730 square feet on the bottom, was completed in 1871. In 1873, however, the percolation of sewage from cesspools and other sources on the surface already affected the well so seriously that the rate of pumping had to be considerably reduced; and, notwithstanding its depth, it has been impossible to draw from it anything like the quantity of water that it yielded at the outset.

The quality of the water obtained from deep wells is on the whole similar to that yielded by springs in corresponding formations; but, for the reasons already given, is more liable to pollution than the latter.   It is also generally harder than spring water.

28. Before leaving this part of the subject the property of "hardness" in water may be briefly discussed, as it is not only of sanitary but of economic importance, and is possessed in some degree by all waters.   The hardest are those derived from calcareous and cretaceous rocks, but the Triassic strata also furnish very hard water.   To a certain extent the presence of salts of lime and magnesia in water is beneficial to health, and highly desirable.   Very soft waters containing much free carbonic acid actively attack the lead of pipes and cisterns ; and although the insoluble carbonate of lead, once formed on the surface of the metal, greatly delays the further corrosive action of the water ; still, much is necessarily detached and is delivered to the consumers.   The Loch Katrine water— notably pure, and of less than 2 deg. hardness—acts energetically upon the lead of pipes and cisterns, and has been responsible for certain injurious results.*   The great variation in hardness of waters derived from different sources may be indicated by the following Table :—

TABLE GIVING APPROXIMATE HARDNESS OF VARIOUS WATER-SUPPLIES.

	Source.	Hardness on Clark's Scale.
London : Kent Water Co.	Wells (Chalk)	25°
West Middlesex Co. ..	R. Thames	15°
Liverpool (Vyrnwy)	Catchment (Silurian)	4°
Do. (Rivington)	Do. (Carboniferous) ..	3°
Manchester (Longendale) ..	Do. (Carboniferous) ..	3°
Glasgow	Lake	1°
Edinburgh	Catchment (Silurian)	7°
Sunderland	Wells (dolomite)	30°

* Report of evidence given by Dr. William Wallace and Dr. Robert Bell, of Glasgow, in the Parliamentary enquiry into the St. Mary's Loch Water Scheme for Edinburgh, 1871.

Whilst very soft waters are in some respects objectionable, the same may be said of those that are extremely hard; and modern practice is tending to reduce, where possible, supplies for domestic and manufacturing purposes to about 5° (Clark's scale). In this connection it is highly important to observe both the "permanent" and the "temporary" hardness of water. Ordinarily, their sum only is stated, unless definitely required from the analyst; and it may be observed that carbonates and sulphates of potash and soda, and many other alkaline salts, produce a softening effect upon water. We shall return later to the subject of artificially softening water.

29. In considering the merits of any given source of water-supply for dietetic purposes, the three principal elements of the question are the quality and the quantity of the water available, and the cost of obtaining it. An exact determination of these elements is, in every case, a matter that involves both labour and skill; and when, as usually happens, the engineer is called upon to decide the relative merits of several suitable schemes, his judgment and experience are greatly exercised to narrow the issue, in the first place, to the factor of cost. In the investigation of water-supply for power or irrigation purposes, the quality is of less vital importance—the principal elements being those of quantity, available fall, and cost.

30. As appears from the considerations already stated, the quality of the water yielded by every source of supply, depends entirely upon the nature and extent of the polluting influences to which it has been subjected, from the time of its condensation from aërial vapour to that of its arrival at the place where it is intercepted in its course back to the ocean. A close examination of the district which forms the catchment-area of the source in question, the sanitary conditions that obtain there, the class of agri-

culture and other industry practised in it, its meteorological data, and the complete analysis of the constitution and inhabitants of the water itself, form the basis of fact upon which must be decided the fitness of the supply for dietetic purposes. The question of the quantity available from any source is equally complex. It involves determination of the rainfall in all its aspects, and the geological structure of the district ; together with all the local circumstances that affect " loss " by evaporation, percolation and absorption (§ 55).

For power-purposes, the available quantity and fall are concerned jointly, since those are the co-efficients for hydraulic energy. To ascertain with any reasonable degree of approximation the power-capacity of a source of supply, involves an investigation into the loss of head incurred by the conveyance of the water, which is by no means a simple matter, and will be better understood if developed in the course adopted in the following chapters.

The subject of cost is one hardly within the scope of this treatise. It is so far dependent upon local circumstances, and upon the particular class or style of the work carried out, that no useful purpose can be served by an essay upon the subject ; for, after all, it is one only to be satisfactorily dealt with in the light of practical knowledge of hydraulic engineering construction.

31. Broadly speaking, waterworks may be classed as either " gravitational " or " pumping." To the former class belong all those aqueducts in which flow takes place under the action of gravity alone ; also Artesian and other spring-supplies under pressure. To the latter belong all other supplies derived from rivers, lakes and wells situated so low with reference to the place of distribution of the water, that mechanical agencies must be employed to deliver the water to the consumers at a proper pressure. The relative

levels of the source of supply and the place of distribution will generally be sufficient to decide into which class the works will come ; although cases sometimes occur in which they may be partly gravitational, but where, owing to some intervening high ground, pumping may have to be resorted to. Local circumstances will, in every case, determine the particular kind of pumping machinery most suitable for the work, whether actuated by steam, wind, or water-power.

Should it appear that the source is so favourably situated with respect to the place of distribution, that the force of gravity may be used to convey the water, it is necessary to ascertain definitely the feasibility of the scheme. For this purpose a survey is required, to determine, first, the facilities that exist for storing water, and second, the most favourable route for the conduit or aqueduct. Preliminary reconnaissance of the district may be made with the aid of general linear and geological surveys, accompanied by such levelling as can be done with the aid of the aneroid barometer. But it cannot be too strongly urged that general views founded upon such surveys must be in every particular subsequently checked, and modified where necessary, by exact chaining and levelling, before detailed designs are prepared.

32. In all cases, the use of the aneroid barometer must be accompanied by observations of the thermometer. If this precaution be not taken, the survey may, and in the case of tracing an aqueduct-contour, probably will, be valueless ; as the best instruments, although corrected for their own variation of temperature, do not effect the requisite correction for that of the atmosphere. Aneroid barometers are generally graduated on the assumption that the temperature of the atmosphere is constant during the observations—a condition that hardly ever obtains in practice. It is therefore well, when using the instrument,

to observe and book the actual atmospheric pressure recorded by the instrument, and at every such reading to
observe the simultaneous atmospheric temperature. The
relative levels of any two points of observation are then
deduced by a simple arithmetical process, arrived at
thus :—

First find the height of the homogeneous atmosphere,[*]
H, at the times and places in question. The value of
this quantity at zero Centigrade, in the latitude of London,
is nearly 26,210 feet. If t_a and t_b are the absolute temperatures of the atmosphere, expressed in degrees Centigrade, at two stations designated a and b respectively,
the required value of H at the locality of those stations is
given by the equation

$$H = 26,210 \left(\frac{t_a + t_b}{273 \times 2} \right),$$

assuming that the mean temperature is half the sum of
the observed temperatures, and that the force of gravity
and the hygroscopic state of the air has not sensibly
changed between the times of observation at a and b
respectively.

Having obtained the true value of H, the levels of
stations a and b are thus arrived at—

Let P denote observed atmospheric pressure ;
    x „   heights in feet measured up from a datum
          plane ;
    ρ „   density of the air at the point of observa-
          tion ;
    g „   force of gravity in kinetic units.

[*] The height of the "homogeneous atmosphere" at any place is the
height of a homogeneous column of air, of uniform cross-sectional area,
required to produce upon its base, under the action of gravity, the actual
atmospheric pressure observed at the place in question—the density of the air
throughout the entire height of the column being supposed to be constant.

Then
$$P = \rho g \cdot H \; ;$$

also
$$- dP = \rho g \cdot dx = \frac{P}{H} dx.$$

Denoting by suffixes the values of P and x at the several stations considered,

$$\int_{P_a}^{P_b} - \frac{dP}{P} = \int_{x_a}^{x_b} \frac{dx}{H} \; ;$$

therefore
$$\log \frac{P_a}{P_b} = \frac{x_b - x_a}{H},$$

whence
$$x_b - x_a = H \left(\log P_a - \log P_b \right) \; ;$$

and, converting hyperbolic into common logarithms, and substituting the value of H at, say, London, at $0°$ C.,

$$x_b - x_a = 60{,}350 \left(\log_{10} P_a - \log_{10} P_b \right).$$

Proper reduction of the levels according to this method is absolutely necessary in aneroid surveying ; and, although we do not propose to describe the use of other instruments used in engineering surveys, we consider this digression justified by the extreme importance of the subject, and the absence of practical explanations relating to it from the hand-books ordinarily consulted by engineering surveyors.

In illustration of the matter, it may be observed that a good aneroid barometer kept at an altitude of, say, 1000 feet above sea-level, may, in the course of an ordinary summer's day, indicate on its dial an oscillation of level of 30 feet—a quantity more than sufficient to nullify contouring observations or attempts to trace a conduit route. Again, the error introduced into the measurement of an absolute height of only 1000 feet from the sea-level may, on a warm day, if the dial of the instrument simply be taken as correct, amount to 90 feet.

33. With regard to the selection of waterworks sites and routes, it would be impossible to formulate rules with-

out premising a knowledge of the constructive and other
principles to which the remainder of this work is devoted;
in addition to an acquaintance with the costs and detailed
methods of carrying on such operations, which is to be
gained through experience alone.

Having sketched the leading considerations involved
in the selection of a source of water-supply, and its fitness
so far as quality is concerned, we pass naturally to the
further question—What quantity of water may the source
be relied upon to yield? The estimation of this quantity
is arrived at by a process that, of itself, requires the aid of
an almost distinct science, forming one of the most important branches of hydraulics; and, as the methods involved
in it are applicable to several necessary proceedings in
waterworks engineering, the subject will be treated separately in Chapter II.

34. In concluding the present section, allusion may be
made to the question of compensation, payable in water,
for interference with the natural order of things in a
gathering-ground that is appropriated to purposes of
water-supply. It is needless to discuss here the well-worn
subject of damage resulting from the abstraction of water
from streams, or of benefits accruing to riparian pro-
prietors from the establishment of impounding-reservoirs,
the mitigation of floods and the equalisation of the flow
in the watercourses of any given district. Into the legal
aspects of the question, it is not our place to intrude.
But, as a matter of customary practice, it may be observed
that the amount of compensation-water allotted to a
stream of which the entire catchment-area is requisitioned
to supply the impounding-reservoir, has been frequently
fixed at one-third of the average yield of the gathering-
ground in question. The compensation-water delivered by
the Liverpool Waterworks to the River Vyrnwy, amount-

ing to 4930 million gallons per annum, is only one-fourth of the available yield of the catchment-area ; the industrial interests affected by diminution of the normal flow of that river are comparatively unimportant—the preservation of the salmon-fishery being the principal object for which compensation-water was required. This contrasts strongly with the onerous conditions imposed upon Liverpool in 1847, when one-half the available yield of the Rivington catchment-area was required to be delivered as compensation; subsequently, however, in 1868, it was reduced by purchase to the present quantity of about one-third of the available yield.

The early experience of Manchester was somewhat similar to this. By an Act of Parliament obtained in 1848, the Corporation was obliged to deliver two-fifths of the available yield of their Longendale watershed as compensation to the River Etherow ; but, in 1854, an Act was obtained enabling this quantity to be reduced by purchase to about one-third of the yield—and this in an important manufacturing district. In 1879, the Manchester Corporation obtained Parliamentary powers to construct the Thirlmere Waterworks, the compensation-water required in respect of the gathering-ground thus appropriated being hardly more than 2000 million gallons per annum—about one-tenth of the available annual yield of the gathering-ground.

The principle of compensating streams affected by the withdrawal of water for any purpose, is undoubtedly proper, and hardly to be fully met, except in special cases, by pecuniary payments. The portion of the flow of an ordinary stream available for useful purposes, without arti-ficial storage, was, however, formerly much over-estimated ; and modern practice in this country, authorised by recent legislation, distinctly recognises that fact.

D

CHAPTER II.

THE MEASUREMENT OF WATER.

35. HYDRAULIC engineering practice necessitates the gauging of water both in the several modes of its occurrence in nature, and in the course of its flow through artificial conduits of various kinds, under the following conditions :—

(1) As it exists in the atmosphere in the form of vapour, and as it falls from the clouds in rain. (2) In streams; in rivers, formed by the confluence of streams, or by water draining from the land into their beds ; and in artificial channels and pipes. (3) By weirs, notches, and orifices, for purposes of irrigation, compensation and general supply ; and by meters, for industrial or domestic supply. We proceed to consider, in detail, some of the most approved methods adopted to meet these requirements.

36. (1) It need scarcely be pointed out here that, in addition to the visible supply of water to every catchment-area in the form of rain, and its escape therefrom by flow over and through the earth, there are continually in progress vast operations of supply and abstraction of water, due to natural condensation and evaporation over the entire surface.

Hygrometric observations are essential to indicate the supply to watersheds in the form of dew, and the loss therefrom by evaporation ; for the most potent factor affecting these phenomena is the extent to which the atmosphere is charged with moisture at any given time.

37. Every circumstance that reduces the temperature of a body of air containing moisture, effects, at a certain temperature, a deposition of dew from it; and this temperature is called the "dew point."

Attention must, however, be drawn to the fact that much of the dew and frost deposited upon plants consists of moisture exhaled from the plants themselves, and emanating from the moist earth.* The quantity of aqueous vapour present in the atmosphere depends, at any given temperature, upon the amount of evaporation and deposition that has previously taken place ; and for every temperature there is a definite maximum quantity of moisture which the atmosphere is capable of containing as vapour. Air in which such quantity is present is termed "saturated."

Evaporation, if the air be not saturated, is more active at high than at low temperatures, and decreases as the quantity of aqueous vapour in suspension increases towards the saturation limit. It is also greater in a dry wind, as fresh bodies of air are thereby continually supplied to absorb more moisture. Evaporation ceases when the air is saturated with moisture.

38. Atmospheric air and aqueous vapour forming a mechanical mixture, the pressure or tension of each constituent is the same as if the other were absent;† and the resultant pressure is their sum. Therefore, the tension of the aqueous vapour in any sample of air, is a measure of the quantity present. This is called the "absolute humidity."

The "relative humidity," or simply the humidity, is the

---

* Quarterly Journal Met. Soc., vol. xvii., 1891, pp. 80 *et seq.*
† The presence of aqueous vapour diminishes, however, the *density* of air.

ratio of the quantity of aqueous vapour present in any portion of the atmosphere to the quantity required to saturate the air at the temperature then obtaining.

39. Various instruments are used for measuring humidity; the best known forms being the Regnault, the Daniell, and the "wet-bulb" hygrometer. The first mentioned is in many respects the best. The Regnault hygrometer consists of a frame carrying two thin polished silver thimbles, about $\frac{3}{4}$ inch in diameter and $1\frac{3}{4}$ inch high, into which fit the lower ends of small glass tubes, the upper ends being provided with corks. Thermometers inserted through the corks pass almost to the bottom of both thimbles. A small pipe passes through the cork to the bottom of one of the tubes, which is half-filled with ether; and by means of an aspirator attached to a second pipe passing through the cork, a current of air is drawn through the ether. By the evaporation of the ether thus produced, the temperature of the tube in question is gradually reduced. The temperature at which dew is deposited on the external silver thimble is noted, and the action of the aspirator is stopped when this occurs; the temperature at which the dew disappears is also observed, the mean of the two being taken as the dew-point. The thermometer in the other tube indicates the temperature of the atmosphere. The polished silver thimbles prevent radiation of heat from the body of the observer to the thermometers, and facilitate the observation of the deposit of dew. The tension f, corresponding with the temperature of the dew-point, is found from a table of tensions of aqueous vapour, as is also the tension F corresponding with the temperature of the atmosphere :

Then $\dfrac{f}{F}$ is the humidity of the atmosphere at the place and time of the observations.

The tension of aqueous vapour was experimentally determined with great accuracy by Regnault, from whose tables the following values are abstracted.*

TENSIONS OF AQUEOUS VAPOUR BETWEEN − 10° C. AND 34° C. EXPRESSED IN MILLIMETRES OF MERCURY.

Temperature.	Tension.	Temperature.	Tension.	Temperature.	Tension.
° C.	mm.	° C.	mm.	° C.	mm.
− 10	2·078	5	6·534	20	17·391
− 9	2·261	6	6·998	21	18·495
− 8	2·456	7	7·492	22	19·659
− 7	2·666	8	8·017	23	20·888
− 6	2·890	9	8·574	24	22·184
− 5	3·131	10	9·165	25	23·550
− 4	3·387	11	9·792	26	24·988
− 3	3·662	12	10·457	27	26·505
− 2	3·955	13	11·162	28	28·101
− 1	4·267	14	11·906	29	29·782
0	4·600	15	12·699	30	31·548
1	4·940	16	13·535	31	33·405
2	5·302	17	14·421	32	35·359
3	5·687	18	15·357	33	37·410
4	6·097	19	16·346	34	39·565

Descriptions of the construction and use of other kinds of hygrometers may be found in text-books of physics.

40. The variation of the humidity is somewhat masked by the diurnal cycle of change which it undergoes. Thus, on land removed from the sea and not influenced by wind, the absolute humidity is greatest in the afternoon, when the temperature of the ground is a maximum ; and least when the temperature is a minimum, shortly before sunrise. But, on the shores of an ocean, the sea-breezes by

* Regnault, 'Relation des Expériences des Machines à Feu,' tome i. pp. 627 *et seq.*—Paris, 1847.

day affect the conditions ; and a secondary minimum absolute humidity obtains during the hottest part of the day, due to the sea-breeze induced by the warm air rising from the ground.

The diurnal variations of humidity are not to be confounded with the daily oscillations of the barometer, which are due principally to the upward and downward currents of air heated near the earth's surface by the sun's rays, and cooled by radiation. These primary motions are temporarily resisted by the inertia and viscosity of the atmosphere, which intensify the variation due to the former cause.

41. From the records of the humidity inferences may be drawn as to the amount of dew deposited, and absorbed for the requirements of vegetable growth. An idea may also be formed of the amount of evaporation, which is especially important in hot countries, where it is considerable unless the humidity is high. All kinds of ground, in particular chalk, fine loam, or sand, possess the property of absorbing by capillary action moisture condensed from the atmosphere, and the humidity of the latter is the determining factor by which such a process must be estimated.

The accession to the rainfall of water in the form of dew is ordinarily neglected ; and the absolute yield of water from a catchment-area is stated as the difference between the quantity of rain falling upon it and the amount of water lost by evaporation, absorbed by vegetable growth, and percolating into the deep-seated rocks of the earth's crust.

The rainfall is estimated from direct measurements. The losses due to the three causes mentioned can only be estimated indirectly by comparison with similar localities under similar conditions ; and, in instituting such comparisons in respect of loss by evaporation and vegetable absorption, the questions of humidity and temperature should receive due attention.

42. Whether supplies of water are to be drawn from catchment-areas, rivers, springs or wells, the estimation of the rainfall upon the area from which the water it is desired to intercept and take is derived, forms the basis of investigation into the capabilities of those sources.

43. The rainfall of a district varies with its situation, physical configuration and altitude, and the direction of the prevailing winds. Where the prevailing winds are warm, and heavily charged with moisture by crossing a large extent of ocean, the rainfall of the first high ground encountered by them will be heavy. The moist air, rising to the altitude of the hills, expands in volume and is reduced in temperature, in accordance with the adiabatic law for the expansion of gases and vapours. The cooled air cannot hold in suspension so large a quantity of vapour as before and the latter is deposited in the form of mist, rain, hail or snow. The rainfall of a district is likely to be small if the prevailing winds traverse a wide expanse of land before reaching it, or if they come from a place of low temperature to a warmer district of no greater elevation. Under such circumstances the air is generally in a suitable state for absorbing additional moisture.

The west coast of Scotland, the north-west coast of England and the greater part of Wales afford illustration of the first condition. The westerly winds, laden with vapour derived from the Atlantic Ocean, encounter broadside the high land near the coast ; and precipitation of the moisture carried by them takes place mainly in that locality. The average annual rainfall in these districts is frequently found to exceed 80 inches ; the maximum being at Seathwaite, in Cumberland, where the rainfall averages about 150 inches per annum.

The average annual rainfall on the west coast of India, where similar conditions prevail, has been known to attain 250 inches, and on the Khasi Hills it is almost double that amount. The flat districts east of the elevated tract of country running from north to south of England, afford examples of the second condition. The westerly breezes, having deposited their burden of moisture on the high land referred to, and becoming warmer as they descend on its eastern side, are no longer capable of depositing moisture in any large quantity; and the average rainfall over a considerable area on that side of the country hardly exceeds 20 inches annually.

The yearly rainfall of every district is subject to considerable variation, the rainfall in a wet year being often more than double that in a dry year. The fluctuation of the daily rainfall is still more marked.

A fall of nearly 7 inches of rain in 24 hours has been recorded at Seathwaite. At some places on the shores of the Mediterranean Sea, in India, and in other parts of the world, a rainfall exceeding 30 inches in a day has been experienced; and in the 24 hours ending 9 A.M., 3rd February, 1893, rain fell to the extent of 35·71 inches at the observatory, Crohamhurst, on the D'Aguilar Range, South Eastern Queensland.*

The torrential discharges which accompany the typhoons of the eastern seas are for a brief period of time probably still more excessive than those just mentioned.

44. Whilst any portion of the earth's surface is gaining heat the rainfall of that district is generally not great; for the atmosphere as a whole, if not disturbed by persistent wind, is rising in temperature and is therefore absorbing moisture. But after that period is passed and the average temperature of the atmosphere is reduced, it cannot hold

---

* *Nature*, vol. xlviii. p. 3.

as much moisture as before. A fall of rain then occurs if aqueous vapour is present in sufficient quantity. The prevailing winds, however, are usually the chief regulators of the time and amount of the rainfall.

45. It sometimes happens in this country that, over limited areas, falls of rain of, say, $\frac{1}{2}$ inch may occur at the rate of 6 to 7 inches per hour ; whilst a fall of, perhaps, 2 inches might occur at the less rate of 2 to 3 inches per hour. A fall of one inch in $\frac{1}{4}$ hour is by no means a rare occurrence in Great Britain. The following statement shows the extremes of rainfall attained in the British Isles.[*]

0·55 inch in 5 minutes.	1·50 inch in 45 minutes.
1·10 „ „ 15 „	1·80 „ „ 60 „
1·25 „ „ 30 „	2·20 „ „ 120 „

These heavy falls of rain are of great interest and importance, as they are the determining causes of the excessive floods that occur on catchment-areas ; and, as might be supposed, the relative magnitude of such floods is greater in the smaller areas.

During the construction of the Woodhead reservoir of the Manchester waterworks, a flood occurred on the 7th October, 1849, which caused, temporarily, discharge at the rate of 500 cubic feet per second per 1000 acres, off a catchment-area of 7500 acres, equivalent to rainfall at the rate of 12 inches per 24 hours over that area.[†] At a later period of the construction of the same works, when the Torside and Rhodes Wood reservoirs were ready for impounding water, rainfall occurred on the 4th February, 1852, which caused a maximum rate of discharge of 250 cubic feet per second per 1000 acres from the catchment-area of 15,400 acres.[‡] At the Vyrnwy works of the Liverpool Corporation,

[*] Symons, 'British Rainfall, 1884,' p. 128.
[†] Bateman, ' History of the Manchester Waterworks,' p. 154—Manchester and London, 1884. [‡] Bateman, *op. cit.,* pp. 160 *et seq.*

a flood occurred on the 29th January, 1883, which caused a flow, for a short time, at the rate of 177 cubic feet per second per 1000 acres off a catchment-area of 18,000 acres, equivalent to rainfall at the rate of 4·2 inches per 24 hours over the entire area. These floods are among the largest recorded in their respective watersheds.

46. There are two reasons for the decrease of the rate of flood-discharge as the catchment-area increases : (1) Extremely heavy falls only last for a short time, and rain falling in the remote portions of a large watershed takes appreciably longer to flow to the place of discharge than does the rain precipitated at more central parts; so the duration of the flood is prolonged, whilst its intensity is diminished. (2) Heavy falls of rain, occurring only locally over limited areas, naturally affect but slightly the discharge from extensive watersheds.

It is useful to remember that 1 inch of rainfall per 24 hours over 1000 acres is approximately equivalent to 42 cubic feet per second. Also that a fall at the rate of 1 inch per hour corresponds with a discharge of 1 cubic foot per second off an area of 1 statute acre. Great discrimination must be exercised in arguing the probable maximum flood of a catchment-area from statistics relating to its mean * rainfall.

47. After the construction of a reservoir, the maximum rate of flood-discharge from the catchment-area supplying it is generally reduced ; for, whilst the period of maximum flood is short, even if the reservoir be full its surface has to be further raised to the height of the free level of the overflow when the flood attains its maximum intensity. The greatest daily rainfall recorded in the British Isles during 30 years occurred at Seathwaite on the 8th of May, 1884,

* The word "mean" in this connection has the significance of annual average over a long period.

when it amounted to 6·78 inches.* The maximum daily fall is found to occur in wet districts, and rarely exceeds 5 per cent. of the mean rainfall ; whilst the daily fall which bears the highest ratio to the mean rainfall occurs in drier districts, and amounts to as much as 15 per cent. of the mean.

It has been conjectured from data afforded by observations in all parts of this country, that in districts where the mean rainfall is small, the greater portion of it generally falls during the summer months, whilst in districts of greater rainfall the period of maximum intensity occurs later in the year. The period of minimum fall occurs also later in the year in wet than in dry districts, but by a shorter time.

48. The rainfall of any given district is estimated from the results of measurements by rain-gauges, described hereafter. These instruments should be placed on flat open ground, with their rims horizontal and elevated about 1 foot above the surface of the ground, so as to prevent rain from splashing into them. They must be situated in positions unsheltered by trees, buildings or banks, and must not be placed near to the crests of steep hill-sides or other abrupt declivities.

The number of rain-gauges required for the estimation of the rainfall of a given catchment-area, depends upon the variation of the rainfall at different parts of the district. For instance, if the rainfall varied only with the altitude of different parts of the district, the number required in the portion situated between any two contours would be proportional to the area embraced by them. But if, as frequently happens, the rainfall also varies at different points of a district which are at the same level, these conditions must be met by duly disposing the rain-gauges in accordance with such variation—an operation which can

* Symons, 'British Rainfall, 1884,' p. 145.

at best be only approximately performed, and demands much experience and skill in such work to ensure any reliable estimate being based upon the records obtained.

49. Rain-gauges should be simple in construction and operation, because they are often necessarily situated in remote places, where trained observers are not available ; and it is desirable that they should be uniform in pattern.

The most suitable instrument for ordinary work consists of a cylindrical copper vessel, about 30 inches long and 6 inches or 8 inches in diameter—the "Snowdon" pattern recommended by the Royal Meteorological Society has a 6-inch cylinder.* Upon this stands a funnel, the rim of which rises vertically about 6 inches, and terminates in a sharp bevelled edge, *Fig. 6.* The funnel terminates in a long straight tube extending nearly to the bottom of a collecting vessel which stands on the base of the cylinder. A small air-space is left between the lip of the collecting vessel and the body of the funnel above it, otherwise water might stand in the funnel and be subject to evaporation.

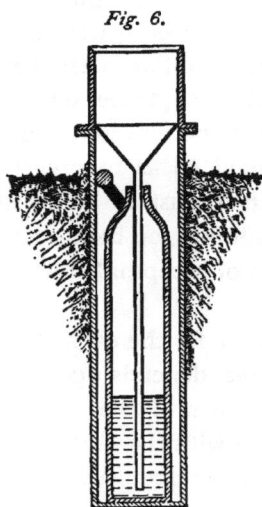

Fig. 6.

Much attention must be paid to the constructional details of even this simple apparatus, in order to ensure accurate results ; and to no part of it does this apply with greater force than to the funnel and the rim. It is recognised that the rim of a rain-gauge serves three objects ; by its rigidity

* Marriott, 'Hints to Meteorological Observers,' published by authority of the Council Royal Met. Soc.—London, 1892.

to maintain the funnel in shape ; to define the area within which the rain-drops are collected ; and to cut any
rain-drop so that the proper portion of it shall pass
into the funnel, and the remainder fall outside.* The
best material of which to form the funnel and rim is probably copper—glass, japanned tin and earthenware tend
to retard the passage of drops of water into the collecting-
vessel, and so permit evaporation to act unduly upon
them. The variation of the area enclosed by the rim, due to
changes of temperature, is so inconsiderable in this country
that it is usually neglected. The best form of edge is that
shown in the diagram, bevelled on the inner side ; an outer
bevelled edge presents greater liability to error due to the
action of wind upon drops falling on the bevelled portion.

The diameter of the funnel, if more than 3 inches, does
not affect the register, but it should be steep internally, to
cause the water to quickly trickle down into the vessel
below. The rain-water collected in the gauge is periodically poured out into a measuring vessel, graduated so as
to show in inches and hundredths of an inch, the amount
which has fallen upon the area of the funnel since the time
it was last emptied.

50. Simpler forms of rain-gauges are successfully employed where they are under frequent and skilled observation ; but experience teaches the hydraulic engineer to use
extreme caution in relying upon the results of amateur
rainfall observations, when the latter are uncorroborated
by other evidence. The fact that, in a well-constructed
rain-gauge, the monthly register exhibits no deficiency
compared with the aggregate of daily readings, is sufficient
to show that its record is, for practical purposes, independent
of the effects of evaporation.

51. The Stutter differentiating rain-gauge has a receiv-

* Quarterly Journal Met. Soc., vol. xvii. p. 136.

ing-funnel which delivers into a smaller funnel furnished with a bent tube, the latter being carried round a vertical axis by clockwork, so as to remain for exactly one hour over each one of 24 fixed equidistant vessels, arranged in a circle. This instrument, therefore, measures the rainfall during each hour of the day.

Automatic recording rain-gauges, although not widely used, are conveniently employed in some circumstances. The Beckley " pluviograph " consists of a receiving-funnel of the ordinary kind, carried by and communicating with a floating vessel, to which a pencil is attached, and which sinks according to the quantity of rain that it receives. The pencil traces a curve on a sheet of paper stretched round a drum rotated by clockwork. When full, the vessel is emptied by a siphon, the pencil rises again to the zero level, and the register is recommenced. Other ingenious forms of recording rain-gauges might be noticed, but the limited application of such apparatus is a sufficient reason for not pursuing the subject further in this place.

52. The rainfall as measured by rain-gauges is apparently greater at the surface of the ground than at higher levels. The diminished record depends largely upon the exposure of the rain-gauge. When the instruments are unsheltered from the wind, they receive less rain than they would if well protected, because the air-currents blowing against the funnels are deflected upwards, and divert some of the raindrops which would naturally fall into the gauge, causing them to fall outside it.

A rain-gauge situated 10 feet above the ground at the Boston reservoir, Rotherham, registered, as the result of 8 years' observations, 9 per cent. less rainfall than that measured at the ground-level, whilst at 25 feet above the ground the record was 12 per cent. less.[*] It appears that

* Symons, ' British Rainfall, 1876,' p. 35.

the decrease of rain intercepted after a height of about 10 feet is attained, becomes comparatively trifling; the reason being that in the latter circumstances there is less interference by the earth's surface with the motion of the wind. The apparent decrease in the rainfall up to about 25 feet above the ground depends upon the exposure of the locality, and is more considerable the greater the force of the wind; also, the snow not collected in elevated rain-gauges reduces their record. Again, the heavier the showers of rain the less is the variation of the record at different levels, because the wind exercises less influence over the motion of a large drop than over that of a small one. In the absence of wind, there appears to be no difference between the rain-gauge records at various elevations above the surface of the ground. No attempt will be made to discuss the variation of rainfall, real and apparent, that obtains at different levels in hilly country. With the information at our command it is impossible to differentiate the element of elevation from the numerous other factors to the joint action of which the results observed in such localities are due.

53. The distribution of rain-gauges throughout most civilised countries is now so extensive that the rainfall of almost any required district may be arrived at directly with considerable accuracy. If the records available do not refer precisely to the locality under consideration, the rainfall of the latter may be arrived at in the following manner. Rain-gauges may be established in the district in question in accordance with the principles already laid down, observations being simultaneously made with instruments previously existing in the neighbourhood. The ratio of the rainfall of the district, thus determined for as long a time as practicable, to the rainfall for the same period registered by the previously-existing gauges, may

be found ; and its application to the rainfall observed by
the aid of the latter in former years, then furnishes
approximately the information required for this district
during those years.

54. For purposes of exact calculation it is necessary
to consider the rainfall during shorter periods than one year,
and it is generally desirable to introduce the average
monthly rainfall into computations relating to the storage
and supply of water. When rainfall records are systematically kept, the register always shows these monthly
quantities.

The following Table, kindly furnished by the Astro-
nomer-Royal, gives the maximum, mean and minimum
monthly rainfall averages at the Royal Observatory,
Greenwich, for fifty years.

MAXIMUM AND MINIMUM MONTHLY RAINFALL, 1841–1891 ; AND MEAN
MONTHLY RAINFALL FOR THE FIFTY YEARS, 1841–1890.

Month.	1841–1891.				Mean 1841–1890 (50 years).
	Maximum.	Year.	Minimum.	Year.	
	inches.		inches.		inches.
January	4·35	1877	0·26	1880	1·99
February.. ..	4·03	1866	0·05	1891	1·48
March	4·05	1851	0·17	1852	1·46
April	4·31	1878	0·09	1855	1·66
May	4·37	1865	0·30	1844	2·00
June..	5·80	1860	0·30	1849	2·02
July	6·75	1888	0·27	1864	2·47
August	5·38	1878	0·45	1849	2·35
September ..	4·12	1871	0·16	1865	2·25
October	7·65	1880	0·76	1879	2·81
November ..	6·00	1852	0·42	1867	2·27
December ..	5·76	1876	0·31	1873	1·77

From this it appears that the average monthly percentage of the mean rainfall at Greenwich is :—

Jan.	*Feb.*	*Mar.*	*Apr.*	*May.*	*June.*	*July.*	*Aug.*	*Sept.*	*Oct.*	*Nov.*	*Dec.*
8,	6,	6,	7,	8,	8,	10,	10,	9,	12,	9,	7 ;

on the Pentland Hills, in Scotland, the average of a shorter, though still considerable, period is :—

11,   8,   6,   5,   6,   6,   7,   8,   9,   12,   12,   10 ;

on the high ground at the southern extremity of the Pennine chain of hills, in England, the average has been ascertained to be :—

9,   8,   6,   6,   5,   8,   11,   10,   10,   10,   9,   8 ;

and on the Snowdon range, in Wales, it has been found to approach :—

7,   9,   6,   6,   7,   6,   8,   9,   8,   10,   12,   12.

These figures, although relating to districts of limited size, widely separated, and subject to very different amounts of rainfall, exemplify generally the average distribution of rain throughout the year in this country ; and, whilst they cannot pretend to express with absolute accuracy anything beyond the particulars recorded by them, they may in the absence of more exact information be useful to indicate generally the proportionate quantities of rainfall which may be expected at the several seasons of the year.

55. To arrive at the actual quantity of water on a catchment-area, which can be rendered available for purposes of supply, in addition to the rainfall three important considerations are involved :—(1) the quantity of rainfall lost through evaporation and absorption by vegetation ; (2) the amount of percolation through the strata out of the catchment-area ; and (3) the maximum period during which the available supply falls short of the demand.

E

56. (1) The portion of the rainfall which finds its way into the streams or springs draining a catchment-area that is but slightly permeable, depends largely upon the activity and extent of evaporation and the absorption of water by plant life. The subject of evaporation has already been alluded to ; its effect is naturally more marked in flat than in hilly districts from which water rapidly flows away, and is in a peculiar degree dependent upon the distribution of the rainfall throughout the year. In valleys with steep hills on each side, subject to a heavy rainfall, the loss from evaporation in this country during the winter is insignificant. In such localities, within our experience, the evaporation and absorption throughout the year have together not exceeded 20 inches, and have been as low as 12 inches ; the loss being experienced principally in the summer months, when it has occasionally amounted to more than 75 per cent. of the rainfall. In dry districts, it frequently happens that during the summer months nearly the whole of the rainfall thus disappears.

Generally, upon permeable soils or upon steep and impervious land, the loss by evaporation is small. If, however, in permeable soils, the surface of saturation is, owing to the physical features of the locality, situated near to the surface of the ground, evaporation takes place actively under favourable atmospheric conditions.

57. Evaporation from the surface of land cannot be directly measured, but is readily obtained as the difference between the rainfall and the percolation and absorption. Careful experiments prosecuted for long periods have been made to ascertain the amount of water which percolates through different thicknesses of soil and other materials, and the average evaporation from a water-surface. The following are the most important of them :—

The result of 30 years' percolation experiments made at

Nash Mills, from 1853 to 1883, with a mean rainfall of 27·8 inches, showed that 6·5 inches of rain percolated through 3 feet of the soil of the district, and 10·6 inches percolated through 3 feet of broken chalk. In each case the materials were contained in a water-tight cast-iron box of 3 feet cube, covered with grass sod. Of the 10·6 inches which flowed through the chalk, 2·1 inches percolated during the summer months, with a rainfall of 14·1 inches ; and 8·5 inches percolated in the winter with a rainfall of 13·7 inches. But, during a dry period of 3 years (1862-64), with a mean rainfall of 22 inches, only 3·5 inches percolated through the soil and 5·1 inches through the chalk.*

Observations during 14 years, upon percolation through soil and sand, and the evaporation from a water-surface, made at Lee Bridge in the years 1860–73, have been fully described in the 'Minutes of Proceedings of the Institution of Civil Engineers.' The averages over the entire period are :—

Rainfall.	Percolation.		Evaporation.		
	Soil.	Sand.	Soil.	Sand.	Water.
inches. 25·7	inches. 7·6	inches. 21·4	inches. 18·1	inches. 4·3	inches. 20·6

The evaporation from the surface of water exceeded the rainfall in 3 years out of the 14. The small evaporation from sand presents a marked contrast to that from the less permeable soil.†

The results of observations made at Fernielaw, Colinton, from the year 1871 to 1880, by the late Mr. James

* 'Hydro-Mechanics': a series of lectures delivered at the Institution of Civil Engineers, 1884-85, p. 17.
† Minutes of Proceedings Inst. C.E., vol. xlv. pp. 19 *et seq.*

Leslie, gave, with a mean rainfall of 38·1 inches, an average evaporation of 15·4 inches from a water-surface; and at the Glencorse filters of the Edinburgh Corporation, from the year 1857 to 1880, with a mean rainfall of 38·8 inches, the average evaporation from a water-surface was 11·4 inches.* Investigations by Sir John Lawes and Dr. Gilbert † have shown how the resistance opposed by increased thickness of strata retains the rain within the influence of evaporation for a longer period, and causes the effect of the latter to be more sensible. Further, they have corroborated the general rule that the ratio of percolation to rainfall during the winter months is much higher than in summer, and is greater in years of heavy rainfall than in dry years.

These experiments possess much interest in their bearing upon some of the general laws applicable to the subject; but their scale is too limited to fix quantitatively, for ordinary engineering purposes, the laws they indicate.

58. The most satisfactory way to arrive at the amount of loss by evaporation, absorption and percolation, in a catchment-area whence the rainfall flows away mainly as a stream, is by continuous gaugings of the stream for as long a period as possible—at the same time observing the rainfall by rain-gauges. In this manner, the actual quantity of available water during the time that the gaugings have been taken, is directly ascertained. The difference between this and the observed rainfall gives the amount of loss from the several causes mentioned, and may be applied, with due regard to the distribution of the rainfall throughout the year and other local circumstances, to other years and to similar localities.

* Minutes of Proceedings Inst. C.E., vol. lxxi. p. 74.
† *Op. cit.*, vol. cv. pp. 35 *et seq*.

The classic gaugings of the River Lee for the years 1851, 1852, 1856, and 1858 gave the following results :—*

Year.	Rainfall.	Flow off the District.
	inches.	inches.
1851	22·62	6·00
1852 †	39·71	9·13
1856	23·91	5·57
1858 †	20·86	4·39

More recent investigations extended to the northern and western districts of Great Britain, which must be regarded as containing the principal sources of water-supply by catchment in this country, have established the fact that, on the less permeable formations, the average annual loss of rainfall by evaporation, absorption and percolation varies between 10 inches and 20 inches, according to the geological and physical character of the district, and its mean rainfall.

59. The distribution of this loss throughout the year, just as in the case of the rainfall, must be next considered. It can only be ascertained exactly for any given district by prolonged experiments embracing a wide range of meteorological conditions. An approximation may be arrived at, as for rainfall, by the application of a ratio derived from observations in similar districts subject to the same meteorological phenomena. The value of such an investigation depends, however, largely upon the accuracy of judgment employed in framing the analogy upon which the entire fabric of the argument rests.

On the impervious and barren Silurian formations of the

---

* Beardmore, 'Manual of Hydrology,' p. 151—London, 1862 ; Rivers Pollution Commission of 1865, 2nd Report, Minutes of Evidence, p. 35.

† 1852 was a specially wet year over the greater part of the country, and 1858 was an exceptionally dry year.

West of England, the average yield of a catchment-area, expressed as percentage of the mean rainfall, may be found, approximately, to be—

During three months :

Jan., Feb., Mar.	*Apr., May, June.*	*July, Aug., Sep.*	*Oct., Nov., Dec.*
25 per cent.	10 per cent.	15 per cent.	30 per cent.

To illustrate the great difference in this respect which obtains in permeable districts of low mean rainfall, with these figures may be contrasted the yield of the Thames valley as indicated by the discharge of the river at Teddington,[*] which may be taken, for purposes of comparison, as—

During three months :

Jan., Feb., Mar.	*Apr., May, June.*	*July, Aug., Sep.*	*Oct., Nov., Dec.*
13 per cent.	6 per cent.	3 per cent.	9 per cent.

60. (2) The percolation experiments alluded to in § 57 convey some idea of the rate at which underground water is derived from the surface in permeable districts. In localities, however, which are suitable for the collection of surface-water, that which percolates the soil is largely restored to the stream that drains the catchment-area ; and, whilst the formation of an impounding-reservoir undoubtedly raises the surface of saturation in its vicinity, the loss of water by percolation into neighbouring valleys or into the land below the site of the dam, is, in places well adapted for such works, extremely small.

Thus far we have dealt with average hydrological conditions, in considering the several important factors that enter into the estimation of the yield of water from catchment-areas. We have now to approach the question, how far the great variations of those conditions, which

---

[*] 'Report on the Flow of the Thames,' by A. R. Binnie, 1892.

occur in an apparently irrational manner, affect the calculation.

61. (3) The maximum period during which the demand for water exceeds the natural yield, determines the amount of storage required. If this condition never occurs, storage is unnecessary. In supplies from catchment-areas, it is sometimes desirable that all the water that can be economically stored, should be impounded. Owing, however, to the great variation in the rainfall from year to year, it is seldom attempted to construct a reservoir large enough to equalise the supply over a longer period than 3 years ; and usually the driest three consecutive years are considered in this connection.

The following rules, for the British Isles, based upon the observations of a long series of years, may be accepted unreservedly.*

(1) The wettest year has a rainfall nearly half as great again as the mean ; (2) The driest year has one-third less rainfall than the mean ; (3) The driest two consecutive years average each one-fourth less rainfall than the mean ; (4) The driest three consecutive years average each one-fifth less rainfall than the mean.

The extremes of wetness and dryness are less pronounced in districts of great rainfall, than in drier districts, consequently more storage is required in the latter than in the former localities.

62. After ascertaining the rainfall for the driest three years, it is necessary to consider the distribution of the rainfall during those years. For purposes of preliminary investigation, the average monthly ratios or percentages of rainfall (§ 54) for the district under consideration, may be employed to determine as a rough guide the distribution of the rainfall during the dry period. To the figures thus

* Symons, ' British Rainfall, 1883,' p. 32.

arrived at, must be applied the proper values of the ratios of yield to rainfall.

As indicated in § 61, the portion of the total yield which can, without undue expense, be rendered available for supply is frequently estimated upon the " driest three years." basis.   The amount of storage required to effect this, forms a problem that is treated separately in Chapter IV.

If, however, the three dry years considered do not include an exceptionally dry year, a quantity of water must also be impounded equal to the difference (after allowing for the effects of evaporation, &c.) between the average annual rainfall of the driest three years and the rainfall of the driest year.   That is to say, $\frac{1}{3} - \frac{1}{5} = \frac{2}{15}$ of the mean yield must be rendered available, in addition to the storage computed upon the basis of supplying the three dry years' yield.

We have it on the high authority of the late Mr. Thomas Hawksley, that, in this country, to be perfectly safe, the storage required may vary from 150 days' supply in wet districts up to double that quantity in dry districts.*   Between these limits, local circumstances determine the requisite storage.   The method of calculation is further treated in § 190.

63. It will now be apparent that the important questions both of the ultimate "yield," or limiting capacity in respect of furnishing a supply of water, of every catchment-area, and of the works necessary to render supplies drawn from such sources uniform in quantity and free from the extreme fluctuations of the natural supply of water to the earth, hinge upon the accurate estimation of the available rainfall of the district.   This quantity is arrived at in either of two ways :—First, by the processes sketched in the preceding sections, the probable minimum rainfall and the loss therefrom owing to unavoidable natural causes, may be ascertained for a proper period, as

* Minutes of Proceedings Inst. C.E., vol. xxxi. p. 56.

well as its distribution during that period. This plan is generally adopted for preliminary investigation. Second, the actual quantity of water flowing away from the proposed source of supply may be directly measured at the place where it is to be intercepted, during as long a period as possible ; such actual measurement affording, in relation to rainfall statistics, data upon which the probable yield of the source may be computed with great accuracy by simple proportion.

64. We proceed to consider the subject of gauging the flow of water. The motion of water is, as yet, only partly amenable to theoretical calculation ; in fact, it is only one phase of it, and that the least prevalent in nature from an engineering point of view, that is included in the theory here presented.

The theory of hydrodynamics is considered in one of two ways : either (1) the relation between the velocity, pressure and density is found at any particular point within the volume of the fluid, which is called the " Eulerian " method ; or (2) the relation between the velocity, pressure, and density of a particular particle throughout its motion is investigated ; this is the " Lagrangian " method. Both of these tacitly assume that particles of fluid follow each other in stream-lines. The discrepancy between the results thus calculated and certain experimental data, and, on the other hand, the agreement between theoretical and experimental results in other cases, point to the conclusion that the motion of water is not always completely specified by assumptions of laminar or multifilar flow. Experiments have shown that when it is so, the motion of water differs from that implied in the ordinary theory. It may be classed broadly as " steady " and " unsteady," and a good idea of the difference between these two kinds of motion is afforded by a simple experiment, of which the following is a short description. If water be allowed to flow through a glass tube, at first very slowly, and then with gradually increasing velocity, and if colour be injected at the inlet end of the tube, it is seen that, when the velocity of the water is slow, the colour is drawn out in separate and steady stream-lines. This state of things continues until a certain velocity is reached, when the stream-lines break into well-defined eddies. This critical velocity varies with the ratio of the viscosity of the water to its density, and inversely as the diameter of the tube.

65. Among circumstances conducive to steady motion, and consequently to a condition to which the ordinary theory applies, are viscosity, a free surface, converging solid boundaries, and curvature with the velocity greatest at the outside. When any of these conditions are in the ascendant, the motion takes place according to theoretical law ; and this fact, together with experiments with coloured bands, has enabled the foregoing classification to be made. In such cases, integration of the equations of motion leads to a comparatively simple relation of velocity, pressure and density, known as Bernoulli's theorem. This theorem is really a statement of the fact that the energy of all particles of water, due to their velocity, together with the energy due to the pressure and height above a datum plane, is constant along any stream-line.

If v be the velocity of a particle of water in feet per second, p its pressure in pounds per square foot, h its height in feet above the datum plane, H a constant for each stream-line, and w the weight of a cubic foot of water in pounds, the relation is expressed by

$$\frac{v^2}{2g} + \frac{p}{w} + h = H.$$

This equation, together with the principle of the conservation of momentum, leads to the commoner results of theoretical hydraulics. The energy of water may be utilised in any of the several ways implied by the three terms of Bernoulli's theorem. In impulse turbines, the energy of water is utilised in its first, or velocity form ; in water-pressure engines, the energy is utilised in its second, or pressure form ; and in overshot water-wheels it is employed in its third, or head form, by its fall under gravity. In reaction turbines and in undershot water-wheels the energy of water is utilised in two forms : in the former in its pressure and velocity forms, and in the latter in its velocity and head forms. The equation shows why, in a contracting tube, the pressure decreases as the tube contracts ; for, supposing the tube to be horizontal, the energy due to the velocity, together with the energy due to the pressure, is constant ; hence, as the energy due to the velocity increases inversely as the fourth power of the diminishing diameter, the pressure proportionately decreases. This circumstance is the origin of the jet pump ; for it is possible, by sufficiently contracting the tube, to obtain a high vacuum.

The circumstances conducive to unsteady motion have been stated as : variation of velocity across a stream when it flows through still water, solid bounding-walls, divergent solid boundaries, and curvature with the velocity greatest at the inside. For these cases, it is found by experiment that the motion is not that implied by theory ; and experiments with coloured bands have indicated that it is of an eddying nature. An interesting example of the boundary of a fluid causing eddying motion is afforded by the presence of oil on water during a breeze. The thin film of oil spread over the surface of the water introduces one condition of a solid boundary, viz. that of resistance to extension ; and this, although slight, is sufficient to alter the form of instability in the water from waves to eddies below the surface. This also explains why it is possible for the surface of a pond to freeze during the prevalence of a breeze. For, if a very thin skin of ice happens to form during a momentary lull, it introduces the necessary surface condition which causes ripples to cease and eddies to be formed underneath the ice.

66. The flow of water, in almost all kinds of channels, falls under the head of unsteady motion. The fundamental equation of delivery, which expresses Q the rate at which water is conveyed along a channel, in terms of v the mean velocity of the stream and A its sectional area, is

$$Q = v A.$$

The velocity of the water flowing in a natural stream, a conduit or a pipe, is such that the resultant of its gravitational impulse in the direction of its motion is equal to the resistance of its bed and of the eddies within its body. It has been found by experiment that the resistance to the flow of water is proportional to the square of its mean velocity, to the area of the surface of contact, and to its density.* The component of the attraction of gravity parallel to the surface of the stream, equals the weight of the water into the sine of the inclination of the [surface to the horizontal plane. ✓

That is

$$f v^2 p \, \Delta \, l \, w = A \, \Delta \, l \, w \, sin \, i.$$

Where v is the mean velocity in feet per second;

p is the wetted perimeter of the channel-section in feet;

$\Delta \, l$ is the length of a small portion of the channel;

w is the weight of a cubic foot of water;

A is the sectional area of the stream in square feet;

---

* The presence of large quantities of silt in the water, by increasing the specific gravity of the mixture, decreases the velocity of streams.

i is the angle of inclination of the surface of the stream, which equals *sin i* as the angle is always small ;

and f and c are coefficients.

If for $\dfrac{A}{p}$ we write *m*, a quantity that is termed the "hydraulic mean depth," the equation may be written more concisely

$$v = c\sqrt{m\,i}.$$

Thus, the velocity is expressed as the product of a coefficient into the square root of the product of the hydraulic mean depth and the inclination, a formula originally proposed by Chezy, Directeur de l'École des Ponts et Chaussées, in 1775. c is not a constant coefficient, because the mean velocity in a section of the stream (to the square of which the resistance is regarded as proportional) is not the velocity with which the water rubs against the surface in contact with which it flows ; and much of the resistance experienced is due to the eddies in the mass of the water. The formula is therefore rectified to accord with this unsteady motion by varying the coefficient, whose value has been experimentally ascertained by Messrs. Darcy, Bazin and other investigators, for various values of hydraulic mean depth, roughness of the perimeter, and inclination of the surface of the stream.

The expression must be applied with caution, for it frequently happens that, owing to a series of falls in the bed of a stream, its inclination is much greater than that to which its velocity is due; in which case the equation is not applicable. The surface-inclination of small channels of uniform regimen may easily be observed by spirit-levelling ; but in deep and sluggish rivers and canals, the inclination is slight, and requires measurements of great refinement for its accurate determination.

67. The important series of experiments on the flow of water in open conduits begun by Darcy in 1856, and after his death completed by Bazin in 1865, were made in artificial channels of varied shapes and sizes, lined with various materials, and in natural canals.* The conduits were rectangular, trapezoidal, triangular, and semicircular in section. The rectangular sections varied in width from 4 inches to 6 feet 6 inches. The bottom-widths of the trapezoidal conduits were a little more than 3 feet, and the sides of these and of the triangular sections were inclined at an angle of 45° to the vertical. The diameter of the semicircular conduits was a little over 4 feet. The canals varied in bottom-width from about 4 feet to 7 feet. Experiments were made with varying depths of water and surface-inclination, care being taken that the regimen of the flow should be first completely established. The discharge was gauged by causing the water to flow through orifices, the coefficients for which had previously been accurately determined. The velocity of flow varied in different experiments from 8 inches per second to 20 feet per second.

The result of these experiments was to ascertain a series of true coefficients for application in the "Chezy" formula. The coefficients were found to vary with the hydraulic mean depth of the stream, and with the roughness of the material of the surface in contact; and to be independent of the surface inclination.

The variation is expressed thus : †

$$c = \frac{1}{\sqrt{a\left(1 + \frac{\beta}{m}\right)}}.$$

* Darcy and Bazin, ' Recherches hydrauliques '—Paris, 1865.
† *Op. cit.*, pp. 130 *et seq.*

The following values of a and of β apply to four classes of channels experimented upon, of different degrees of roughness :

	Values of a	β
(1) Smooth channels—cement or planed timber	0·0000457	0·10
(2) Smooth channels—ashlar, brickwork or planks ..	0·0000579	0·23
(3) Rough channels—rubble masonry or stone pitching	0·0000732	0·82
(4) Rough channels—earth	0·0000854	4·10

To assist the rapid application of the formula within the limits of the experiments—calling c_1, c_2, c_3, c_4, the respective values of c for the several kinds of channel described, the following table gives their amounts for different hydraulic mean depths (m).

Hydraulic mean depth (m)	Values of c.			
	c_1	c_2	c_3	c_4
0·25	125	95	57	26
0·50	135	110	72	36
0·75	139	116	81	42
1·00	141	119	87	48
1·50	143	122	94	56
2·00	144	124	98	62
3·00	145	126	104	70
4·00	146	128	106	76
5·00	146	128	108	80
6·00	147	129	110	84
7·00	147	129	110	86
8·00	147	130	111	88
9·00	147	130	112	90
10·00	147	130	112	91
15·00	147	130	114	96
20·00	147	131	114	98
∞	148	131	117	108

Applied to artificial channels, and to canals of moderate size, the formula gives satisfactory results ; but it is not generally applicable to large channels, or to rivers, or to torrential streams.

68. In the formula for "c" (§ 67) Messrs. Darcy and Bazin made the coefficient depend upon the dimensions of the channel and its roughness only. The utilisation of more extended data has shown that c also varies with the surface-inclination of the stream. Modified so as to apply to a range of phenomena from those of great rivers, as ascertained by Messrs. Humphreys and Abbott in the Mississippi, to those presented by the torrents of Switzerland, experimented upon by Messrs. Ganguillet and Kutter, in the work of the last mentioned authors [*] the expression for c takes the form

$$c = \frac{z}{1 + \dfrac{x}{\sqrt{m}}},$$

when

$$z = 41 \cdot 6 + \frac{1 \cdot 811}{n} + \frac{0 \cdot 00281}{i}$$

and

$$x = n\left(41 \cdot 6 + \frac{0 \cdot 00281}{i}\right).$$

The expression may therefore be written

$$c = \frac{\sqrt{m}}{n} \cdot \frac{x + 1 \cdot 811}{x + \sqrt{m}}.$$

The values of n for various kinds of channels are :

For surfaces of planed planks,			$n = 0 \cdot 009$
„	„	plaster in neat cement,	„	$0 \cdot 010$
„	„	plaster, $\frac{2}{3}$ sand, $\frac{1}{3}$ cement,	„	$0 \cdot 011$
„	„	unplaned planks,	„	$0 \cdot 012$
„	„	ashlar and brickwork,	„	$0 \cdot 013$
„	„	rubble masonry,	„	$0 \cdot 017$

[*] Ganguillet and Kutter, 'Versuch zur Aufstellung einer neuen allgemeinen Formel für die gleichförmige Bewegung des Wassers in Canälen und Flüssen ' —Berne, 1877. Also translation by R. Hering and J. C. Trautwine—New York and London, 1889.

For canals in firm gravel, $n = 0.020$

For rivers and canals in earth in good order, „ 0.025

„ „ „ in moderate order, } „ 0.030

„ „ „ in bad order, „ 0.035

Additional values of n have been interpolated by Mr. Lowis D'A. Jackson,[*] and others.

69. It may be desirable to observe that it has sometimes been proposed to use the "Chezy" formula with a constant value of c to determine the velocity of streams, but that course can be obviously of only limited application. All the foregoing investigations relate to the determination of the mean or average velocity in a definite cross-section of the stream of water under consideration. This velocity multiplied by the plane area of the cross-section gives the rate of flow on delivery of the stream, which is the quantity actually sought.

70. The motion of water in pipes is generally of a more stable character than that in channels, and the experiments that have been made upon this kind of flow, are sufficient to enable the delivery of pipes to be calculated with a high degree of accuracy, in all cases that occur in ordinary practice. The flow of water in pipes exemplifies both steady and unsteady motion. It was found by Poiseuille that for capillary tubes, the resistance is proportional to the velocity ; and Darcy ascertained that in pipes of larger diameter, the resistance is proportional to the square of the velocity. To bring these apparently inconsistent laws into agreement, Dr. Osborne Reynolds made a series of experiments,[†] proving that the motion of water in all tubes or pipes, at a sufficient distance from the outlet, is steady ; but, after attaining a certain critical velocity which varies

[*] Jackson, ' Hydraulic Manual,' p. 41—London, 1883.
[†] Phil. Trans. Roy. Soc., Part 3, 1883.

F

inversely with the size of the pipes, the motion becomes sinuous.   Also that, whilst the motion is steady, the resistance is proportional to the velocity ; but that, as soon as a sinuous condition occurs, the resistance varies approximately as the square of the velocity.   The fact that the resistance in a capillary tube varies as the velocity, whilst that in a tube of greater diameter varies with the square of the velocity, led to the supposition that some property of water, exerting an important influence on its motion, had not been recognised in the ordinary equations. That the resistance should depend on the absolute size of the tube is impossible, according to the laws of motion. This important factor was found to be the ratio of the viscosity to the density of the water—which ratio has the dimension of the product of a velocity into a space.

Dr. Reynolds was led, from a consideration of his experiments made in 1883, those of Poiseuille made in 1845, and those of Darcy made in 1857, to put the expression for the resistance of the flow of water in pipes into the following form, which expresses in the metrical system the resistance for all diameters of pipes and temperatures :

$$A \frac{d^3}{P^2} i = \left(B \frac{d\,v}{P} \right)^n. \qquad \text{(i.)}$$

Where i is the virtual inclination (§ 71) ;

d is the diameter of the pipe ;

v is the velocity of the water ;

$P = (1 + 0.0337\,T + 0.000221\,T^2)^{-1}$ is the ratio of viscosity of the water to its density (T being temperature in degrees Centigrade) ;

$A = 67,700,000$; $B = 396$;

and n is an index, which is unity when the velocity is below the critical velocity ; and for higher velocities, varies according to the roughness of the internal surface of the pipe.

For clean lead pipes, above the critical velocity, $n = 1 \cdot 79$
,, varnished pipes ,, ,, ,, $1 \cdot 82$
,, clean cast-iron pipes ,, ,, ,, $1 \cdot 88$
,, rusted iron pipes ,, ,, ,, $2 \cdot 00$

Taking the second and the English foot as units, the formula becomes:

$$v = \frac{3 \cdot 28^{\frac{2n-3}{n}} A^{\frac{1}{n}}}{B} \times \frac{d^{\frac{3-n}{n}} i^{\frac{1}{n}}}{P^{\frac{2-n}{n}}}$$

$$= \frac{C}{P^{\frac{2-n}{n}}} d^{\frac{3-n}{n}} i^{\frac{1}{n}}, \; C \text{ being a constant.}$$

When $n = 1 \cdot 88$, the formula becomes

$$v = \frac{60}{P^{\cdot 064}} d^{0 \cdot 6} i^{0 \cdot 53}. \qquad \text{(ii.)}$$

When $n = 2 \cdot 00$, the formula becomes

$$v = 38 \sqrt{d \, i}. \qquad \text{(iii.)}$$

These equations apply to pipes of all diameters, excepting capillary tubes, for the rates of flow that occur in practice. Taking the temperature as $10°$ C., $P^{0 \cdot 064} = \frac{1}{102}$; i. e. the flow through a new cast-iron pipe is 2 per cent. greater at $10°$ C. than at $0°$ C. In the formula for rusted pipes the term P is absent; the flow in such pipes is, therefore, independent of the temperature. Equation (ii.) represents the case of newly laid mains, and equation (iii.) applies to mains fouled by use, and incrusted with nodules of rust.

For facility of working, neglecting effects of temperature, the formulæ may be written thus, taking logarithms,

(ii.) $\log v = 1 \cdot 77815 + 0 \cdot 6 \log d + 0 \cdot 53 \log i.$
(iii.) $\log v = 1 \cdot 57978 + 0 \cdot 5 \log d + 0 \cdot 5 \log i.$

Since *P*, the ratio of the viscosity of water to its density at any given temperature, is of the dimension of a space multiplied by a velocity, it will be seen that, as the index *n* changes from 1 to higher values, the dimensions of the term on the right of equation (i.) remain unchanged.

71. A somewhat similar formula was proposed in 1873,[*] by Dr. C. J. H. Lampe, based upon experiments conducted at Dantsic on a water-main about 16 inches in diameter, and of a length of nearly 9 miles. Dr. Reynolds was led to put the equation into the form (i.) by plotting, from the experiments, the logarithms of the slopes of the pressures as abscissæ and the logarithms of the velocities as ordinates. The factors by which *i* and *v* are multiplied, are the distances through which the resulting straight lines had to be moved horizontally and vertically to bring them into coincidence, and the index *n* is the inclination of the straight lines to the axis of *x*.

In the case of pipes, *i*, the virtual inclination, may be thus explained : If two reservoirs of water at the same level are connected by a pipe, the water in the latter is at rest and the pressure is everywhere equal to that due to the depth of the pipe below the surface of the water in the reservoirs. If, however, the water in one of the reservoirs is lowered, water will begin to flow through the pipe at a velocity dependent upon the inclination of an imaginary line joining the top-water of the reservoirs at the respective ends of pipe, which line is called the virtual gradient. The pressure at any point in the pipe is that due to a head of water equal to the height of the virtual gradient above that point. Thus, if the pipe were tapped anywhere and a tube were carried up from it, the water would rise in the tube till it reached the line of the virtual gradient—the head due to the difference between the level of this point and

[*] *Der Civilingenieur*, vol. xix. pp. 69 *et seq.*

the higher top-water having been utilised in overcoming the resistance of the pipe. If the pipe touched this line at any part, the pressure in the pipe there would be equal to the atmospheric pressure ; and if the pipe rose above the line, the pressure would be less than that of the atmosphere. As, however, it is the practice to place air-cocks at each summit on the line of a water-main, the latter contingency would never occur ; but the flow would be interrupted there, its flow being that due to the gradient between the upper reservoir and that point. Although the effect of temperature, by influencing the viscosity of water, is sufficient, *cæteris paribus*, to double the flow at very low velocities between 5° C. and 35° C. ; at higher velocities its influence is slight, and Darcy does not appear to have noticed it at all.

72. From experiments on the flow of water in pipes varying between $\frac{1}{2}$ inch and 1 foot 8 inches in diameter, Darcy deduced an empirical equation for the velocity of flow in pipes, reduced to the same form as that applicable to streams.* In this, the coefficient is of less fluctuating character ; and although varying for different sizes of pipes, it is, as regards those previously considered for streams, comparatively constant. Hence a mean coefficient may be used for a large range of diameters, without exceeding the limits of accuracy as observed in the experimental data. In the formula $v = c\sqrt{mi}$ for a pipe, $m = \dfrac{d}{4}$, and c varies only with the diameter and the roughness of the internal surface. The coefficient is expressed in the following forms, using English measures :

$$c = \frac{113}{\sqrt{1 + \dfrac{1}{12\,d}}}, \text{ for clean iron pipes,} \qquad \text{(i.)}$$

* Darcy, 'Mouvement de l'eau dans les Tuyaux,' p. 110—Paris, 1857.

and
$$c = \frac{80}{\sqrt{1 + \frac{1}{12\,d}}}, \text{ for rusted pipes.} \qquad \text{(ii.)}$$

If d be taken a mean between 1 foot and 4 feet, the average coefficient so found will yield results within the limits of sensible accuracy. The formula then becomes

For clean iron pipes, $v = 55\sqrt{di}$, or $Q = 43\,d^{\frac{5}{2}}\,i^{\frac{1}{2}}$. (iii.)

" rusted " $v = 39\sqrt{di}$, or $Q = 30\,d^{\frac{5}{2}}\,i^{\frac{1}{2}}$. (iv.)

The similarity between these equations and (ii.) and (iii.) § 70, is worthy of attention.

Experiments made by Mr. J. Leslie on a cast-iron main 15 inches in diameter, at Edinburgh, which had been laid for 30 years and was much rusted, gave a coefficient of 43 for Darcy's formula (ii.) above. With a 16-inch pipe, which had been laid 9 years, and was less fouled, the experiments gave, in three different lengths at different hydraulic gradients, coefficients 50, 48 and 50 respectively in the same formula (ii.).

It may be useful to observe that the loss of head h due to frictional resistance to the flow of a quantity of water Q, through a length l of pipe of diameter d, deduced from the equations (iii.) and (iv.) is

$$h = \frac{l\,Q^2}{1850\,d^5}, \text{ for clean iron pipes,}$$

and
$$h = \frac{l\,Q^2}{900\,d^5}, \text{ for rusted iron pipes.}$$

73. By the formulæ discussed in §§ 66–68, the flow in artificial channels of definite geometrical form and composed of particular materials, may be measured with a high degree of accuracy ; but the considerations adduced show that for natural channels in earth and in rock, a good

* Minutes of Proceedings Inst. C.E., vol. xiv. pp. 273 *et seq.*

deal of uncertainty attaches to the calculations of flow from the data afforded by their dimensions, the surface-inclination and the roughness. This is principally owing to the difficulty of making proper choice of a coefficient suitable to the dimensions, inclination and roughness of the channel in question. Hence, in the case of streams and rivers, where highly accurate results are required, recourse must be had to other methods of gauging.

74. The dry-weather flow of a stream draining an area of, say 20,000 acres, is much less than its rate of flood-discharge. In barren mountainous countries, the former may be as low as $\frac{1}{10}$ cubic foot per second per 1000 acres ; while the latter may occasionally amount to 1000 or even 2000 times the dry-weather flow. The methods of gauging employed in such a case must therefore be adapted to deal with the different states of the flow.

The dry-weather flow of small streams, and those of larger size when there is no traffic and a suitable fall exists in their beds, may be accurately measured by causing the water to pass over a notch-gauge. This apparatus, a cross section of which is given in *Fig. 7*, consists essentially of a wide slot or depression in the upper edge of a vertical plate or board, placed across the stream, normally to its axis. The maximum quantity that can be measured by a notch, depends upon its length and the fall that can be obtained for the water from its up-stream to its down-stream side. On the up-stream side, to prevent the water from approaching the notch with sensible velocity, it is desirable that the channel should be wider and deeper than the notch ; so that the body of water above it may be practically quiescent. If the area of the water, as it passes through the notch, be more than one-fifth part of the area of the channel of approach, the velocity of approach must

Fig. 7.

be taken into account in calculating the discharge over the notch. To avoid sensible velocity of approach, the depth of water below the sill of the notch on the up-stream side should be at least three times the depth of water flowing over the notch ; and the sides of the approach-channel should be distant a like amount from the ends of the notch. By a notch 20 feet long, when a fall of 1 foot in the bed of the stream can be secured, a flow of 20,000,000 gallons per day may be conveniently measured. Larger quantities than this may be gauged by notches ; though for such considerable flows it is generally desirable to resort to current-meter observations, or to the method of gauging by float measurements.

To construct a notch-gauge, a water-tight weir is first made across the stream, and in this is set a frame of wood or metal, on the up-stream side of which are attached thin metal plates, forming the edges of the gauge. The latter need not be bevelled on the down-stream side, if they present a square ¼-inch edge in the vertical plane. In the common rectangular notch-gauge, three plates are used —one of them forming the sill and the others the vertical ends of the notch. The sill is fixed level, and its length must be accurately known. The notch may be provided with movable sides, in order that its width may be reduced when the flow is small ; as it is undesirable that the height of water over it should be less than about 4 inches. The level of the water on the down-stream side must be sufficiently below the sill—not less than one-half the maximum depth of water flowing over it—to allow free access of atmospheric air under the discharging stream.

75. Since the resultant flow in each vertical strip of the notch is

$$d\,l \int_0^{\cdot H} \sqrt{2\,g\,h}\; d\,h,$$

where l is the length of the notch in feet,

h the height in feet of any horizontal filament above the sill,

and H the height in feet of the free-level of the discharging water above the sill,

i. e. integrated, $\dfrac{2\,d\,l}{3}\sqrt{2g}\;H^{\frac{3}{2}}$;

the discharge Q, in cubic feet per second over the notch, if there were no interference by the ends (termed end-contraction), and the edge of the sill with the free flow of the water under the action of gravity, would be given by

$$Q = \frac{2}{3}\,l\sqrt{2g}\;H^{\frac{3}{2}}.$$

Owing, however, to the interference referred to, and to the fact that, when only a small quantity of water is passing, the stream has a tendency to flow down the face of the plate without allowing free circulation of air on the lower side, a coefficient c has to be introduced into the expression. This coefficient is not constant, but varies with different values of l and H. The formula then becomes

$$Q = \frac{2}{3}\,c\,l\sqrt{2g}\;H^{\frac{3}{2}}.$$

To obtain the value of c has been the object of numerous experimenters, among whom may be mentioned Blackwell,[*] Francis,[†] Fteley and Stearns,[‡] Hamilton Smith,[§] Du Buat,[‖]

---

[*] Minutes of Proceedings Inst. C.E., vol. x. p. 331.
[†] Francis, ' Lowell Hydraulic Experiments '—New York, 1883.
[‡] Transactions American Soc. C.E., 1883, pp. 52 *et seq.*
[§] Hamilton Smith, ' Hydraulics '—London and New York, 1886.
[‖] Du Buat, ' Principes d'Hydraulique et de Pyrodynamique '—Paris, 1816.

Bidone,[*] Poncelet and Lebros,[†] Castel,[‡] Boileau,[§] Eytel-
wein[‖] and Weisbach.[¶]

Most of the experiments agree in two particulars :
(1) as the head h becomes less than about 3 inches, the
coefficient c experiences a rapid increase ; and (2) with
values of h between 3 inches and about 2 feet, the coefficient is fairly constant.   Between these limits, and with
a notch longer than 2 feet, in order that the end-contractions may not interfere with each other, Mr. J. B. Francis[**]
proposed a formula which agrees closely with a careful
series of experiments carried out by him, and is extensively
used in hydraulic practice.   This formula, for discharge
from a still pond, is

$$Q = 3 \cdot 33 \, (l - 0 \cdot 1 \, n H) \, H^{\frac{3}{2}} \, ;$$

where n is the number of end-contractions ; i. e. in an ordinary rectangular notch, $n = 2$.   Here c is taken equal to
0·622, and the effective length of the notch is reduced by
0·2 H so as to allow for the reduction in the discharge due
to the influence of the end-contractions.   The experiments,
from the results of which this formula is derived, were made
on a 10 foot rectangular notch, with head varying between
1·6 foot and 0·6 foot, in which range the coefficient varied
less than 1 per cent.   In practical application, the measurement of the height of the free-level of the water above the
sill of the notch must be made at a point far enough away
from it to be unaffected by surface-curvature.

  * Bidone, ' Expériences sur la dépense des réservoirs '—Turin, 1824.
  † Poncelet and Lebros, ' Expériences Hydrauliques '—Paris, 1832 ; Lebros,
' Hydraulique Expérimentale '—Paris, 1849.
  ‡ Mémoires de l'Académie des Sciences de Toulouse, tome iv. p. 238,
1837 ; and D'Aubuisson de Voisins, ' Traité d'Hydraulique '—Paris and
Strasburg, 1840.
  § Boileau, ' Traité de la Mesure des eaux courantes '—Paris, 1854.
  ‖ Abhandlungen der Berliner Akademie, 1814–15 and 1818–19.
  ¶ Weisbach, ' Versuche über den Ausfluss des Wassers '—Leipsic, 1842.
  ** Francis, ' Lowell Hydraulic Experiments,' p. 119—New York, 1883.

76. The following experiments made by us in 1885, by causing the same quantity of water to pass first over a rectangular notch 8 feet long and afterwards over one 17·02 feet long, show the divergence of Francis' formula from the truth as the head falls below 3 inches.

NOTCH 8 FEET LONG.		NOTCH 17·02 FEET LONG.	
Head over sill.	Discharge according to Francis' formula.	Head over sill.	Discharge according to Francis' formula.
Inches.	Cubic feet per second.	Inches.	Cubic feet per second.
1·36	1·01	0·77	0·92
3·45	4·09	2·00	3·86
5·12	7·38	3·03	7·27
6·46	10·36	3·82	10·20
6·99	11·68	4·15	11·62

It will be seen that, whilst the highest three readings agree within $1\frac{1}{2}$ per cent., the lowest two differ by more than 5 and 9 per cent. respectively.

The range of temperature occurring in ordinary practice does not appreciably affect the results of notch-gauging.

Formulæ more accurate than that of Francis, for small depths of water over the notch, have been proposed by a number of experimenters, the valuable work of Messrs. A. Ftelcy and F. P. Stearns demanding special notice.

For rough comparative calculations, the formula for a rectangular over-fall notch may be put into the form

$$Q = \frac{10}{3} l H \sqrt{H},$$

the value of Q being in cubic feet per second and l and H being in feet; or

$$Q' = \frac{l h \sqrt{h}}{2},$$

where the value of Q' is in gallons per second, h is in inches and l is in feet.

It is worth remembering that the number of cubic feet per minute multiplied by 9000 gives an equivalent in gallons per 24 hours.

77. It has always been the endeavour of experimenters to realise the condition of water being discharged over notches from a state of perfect stillness ; and it is desirable that this should be approximated to in all notch-gauging in actual practice, although circumstances sometimes prevent it. When the area of the stream is so small in relation to that of the notch that the velocity of the approaching water becomes sensible, it may be allowed for in the following manner. Call the velocity of approach u, and let h_0 be the head due to that velocity. Then

$$h_0 = \frac{u^2}{2g}.$$

If the velocity of approach were uniform over the area of the notch, the head at each horizontal lamina of water discharged by it would be increased by that amount. Consequently, the discharge would be equivalent to that from a still pond with the head of $H + h_0$ above the sill, less the discharge due to the head h_0; i.e. modifying Francis' formula,

$$Q = 3\cdot33\ (l - 0\cdot2\ H)\left[(H + h_0)^{\frac{3}{2}} - h_0^{\frac{3}{2}}\right].$$

The uniformity of the velocity of approach, over the area of the notch, naturally depends upon the shape of the approach channel. Thus, from a series of experiments made in 1877, Messrs. Fteley and Stearns concluded that h_0 equalled $b\ \dfrac{u^2}{2g}$, where b varied between $1\cdot33$ and $1\cdot87$.[*] For most forms of channel of approach, the latter value would, however, be too high.

Where it happens that the fall in a stream is too slight

[*] Transactions American Soc. C.E., 1883, p. 12.

for an ordinary over-fall notch to be conveniently used, it is sometimes useful to be able to obtain a measurement of the discharge of the stream by means of a " drowned " notch—that is, a notch that so far constricts the stream as to produce a sensible difference of water-level between its upper and lower sides, but in which the water on the down-stream side is above the level of the sill.

In this case, the discharge Q is approximately

$$\frac{13}{4} \, l \left(h + \frac{3}{2} \, h' \right) \sqrt{h} \, ;$$

where h is the difference between the water-levels above and below the notch, and h' is the height of the water-surface on the down-stream side above the sill.

It is frequently desirable to measure roughly the discharge of a stream over a broad-crested weir. From experiments on weirs with crests varying between 2 inches and 10 inches in breadth, it appears that the discharge for heights less than 1·6 times the breadth of the sill is less than for a sharp-edged sill ; when that height is reached, it commences to be greater than for the sharp-edged weir.[*]

For a weir 10 feet long and about 3 feet broad across the sill, the formula, for heads varying between 7 inches and 20 inches, given by Mr. J. B. Francis,[†] is

$$Q = 3 \cdot 017 \, H^{1 \cdot 53}.$$

78. For gauging streams of small and varying discharge, a triangular or V-notch is most suitable.

Fig. 8.

The discharge for any lamina $b \, d \, h$, *Fig. 8*, is

$$b \, dh \sqrt{2gh} \, ;$$

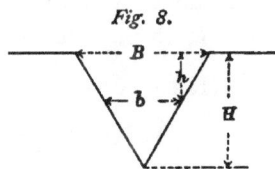

[*] Transactions American Soc. C.E., 1883, pp. 86 *et seq.*
[†] 'Lowell Hydraulic Experiments,' p. 135.

b being the breadth at any depth h, B the top breadth, and H the total depth of flow, the dimensions being in feet.

Therefore, the discharge of the notch is

$$\int_0^H b\sqrt{2gh}\,dh$$

$$= \frac{B}{H}\int_0^H \left(H - h\right)\sqrt{2gh}\,.\,dh.$$

$$= \frac{4}{15} B \sqrt{2gH^3}.$$

Allowing for end-contractions, the formula becomes

$$Q = c\,\frac{4}{15} B \sqrt{2gH^3}.$$

In a right-angled notch, $B = 2H$, and

$$Q = c\,\frac{8}{15} \sqrt{2g}\,H^{\frac{5}{2}}\,; \qquad\qquad \text{(i.)}$$

and in a notch in which $B = 4H$,

$$Q = c\,\frac{16}{15} \sqrt{2g}\,H^{\frac{5}{2}}. \qquad\qquad \text{(ii.)}$$

As the flow is always similar at whatever height the water is discharged through the notch, the coefficient c is much more constant than is the case with the rectangular form. For a right-angled notch with sharp edges, c, as determined by the late Dr. James Thomson, is $0\cdot59$; and for a notch in which $B = 4H$, c is $0\cdot62$.[*]

Hence equation (i.) may be written $Q = 2\cdot54\,H^{\frac{5}{2}}$;

and   (ii.)   „   „ $Q = 5\cdot30\,H^{\frac{5}{2}}$.

79. When it is desired to have a continuous record of the discharge of a notch, a float-gauge is erected—generally in an easily accessible situation on the bank of the

---

[*] British Assoc. Report, 1861—Reports on the State of Science, p. 151.

approach-channel. An inverted siphon-pipe communicates from the still-water pond to the vertical cylinder in which the float works. A wire from the float is attached to a pen or pencil, which records on a drum, rotated uniformly by clockwork, the varying height of the water-surface in the still pond, above the sill of the notch.

80. When the flow to be measured is not great, and plenty of fall is available, gauging by orifices is often both convenient and accurate. A rectangular orifice is the form generally used in gauging ; except for very small quantities of water, when circular orifices are sometimes employed. The discharge, in cubic feet per second from a rectangular orifice in a thin plate, is expressed by the formula

$$Q = c\,l\,\frac{2}{3}\sqrt{2g}\,(h_2^{\frac{3}{2}} - h_1^{\frac{3}{2}}),$$

where l is the length of the slot or orifice in feet ;

$\quad$ h_2 is the height in feet of the surface of water above its lower edge ;

$\quad$ h_1 is the height in feet of the surface of water above its upper edge ;

and c is a coefficient.

For simplicity, it is often expressed in the form

$$Q = c\,l\,d\sqrt{2g\,h},$$

where d is the depth of the orifice in feet, l its length, and h is the height in feet of the surface of the water in the still pond, above the centre of the orifice.

The following table gives the values of c for different dimensions of length and depth, abstracted from the results of numerous experimenters, including Michelotti,[*] Bossut,[†] Poncelet and Lebros, and Hamilton Smith.

---

[*] Michelotti, ‘ Sperimenti Idraulici ’—Turin, 1767.

[†] Bossut, ‘ Traité théorique et expérimental d’Hydrodynamique ’—Paris, 1796.

TABLE OF COEFFICIENTS OF DISCHARGE FOR ORIFICES IN THE
FORMULA $Q = c\,l\,d\,\sqrt{2\,g\,h.}$

Head over centre of orifice.	$l=2$ ft. $d=1$ ft.	$l=2$ ft. $d=6$ in.	$l=1$ ft. $d=1$ ft.	$l=1$ ft. $d=6$ in.	$l=1$ ft. $d=3$ in.	$l=9$ in. $d=9$ in.	$l=6$ in. $d=6$ in.	$l=3$ in. $d=3$ in.
feet.	c.	c.	c.	c.	c.	c.	c.	c.
0·5	0·60	0·61	0·60	0·61	0·63	0·60	0·60	0·60
1	0·60	0·62	0·60	0·62	0·63	0·60	0·60	0·60
2	0·60	0·62	0·60	0·62	0·63	0·60	0·60	0·60
3	0·60	0·62	0·60	0·62	0·63	0·60	0·60	0·60
4	0·60	0·61	0·60	0·61	0·62	0·60	0·60	0·60
5	0·60	0·61	0·60	0·61	0·62	0·60	0·60	0·60
6	0·60	0·61	0·60	0·61	0·62	0·60	0·60	0·60
7	0·60	0·61	0·60	0·61	0·61	0·60	0·60	0·60
8	0·60	0·61	0·60	0·60	0·61	0·60	0·60	0·60
9	0·60	0·61	0·60	0·60	0·60	0·60	0·60	0·60
10	0·60	0·61	0·60	0·60	0·60	0·60	0·60	0·60
20	0·60	0·60	0·60	0·60	0·60	0·60	0·60	0·60

As an application of these formulæ, it is sometimes requisite to calculate the time occupied by a tank or reservoir in emptying through an outlet at the bottom.

Suppose the tank to be of any given form of cross-section, and call its area at any level, A. Let d be the initial height of the water-surface above the outlet, h its height after emptying for any time t, a the area of the outlet, and c the coefficient of discharge.

The discharge through the outlet in any small period of time dt is $c\,a\,\sqrt{2gh}\,.\,dt$.

The volume of the tank emptied in the same time is $A\,d\,h.$

Therefore $c\,a\,\sqrt{2gh}\,dt = A\,d\,h\,;$ (i.)

or $dt = \dfrac{A}{c\,a}\,.\,\dfrac{d\,h}{\sqrt{2gh}}\,.$

For a vertical cylindrical tank, the time of emptying is, by integration, found to be

$$\frac{2A}{ca}\sqrt{\frac{h}{2g}}.$$

For any other form of tank, A is expressed as a function of h, and the equation is integrated as above.

An interesting application of equation (i.) is to find the shape of the vase of the ancient clepsydra, or water-clock. In this case $\frac{dh}{dt}$ is constant; therefore A the area at any height h above the outlet $\infty \sqrt{h}$.

81. For a completely submerged orifice 2 feet long, 6 inches deep, and $\frac{1}{4}$ inch broad, with a difference of water-level between the two sides from $\frac{1}{4}$ foot to $\frac{3}{4}$ foot, we found in a careful series of experiments, made in 1887, that the coefficient c in the formula $Q = cld\sqrt{2gh}$ is 0.67; h being the difference in level of the water-surfaces on the upper A and lower B sides of the orifice, expressed in feet (*Fig. 9*).

Fig. 9.

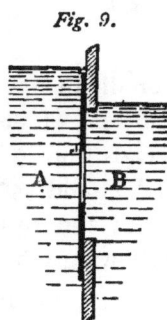

The submerged orifice is important and frequently very convenient, because any variation in the level of the water on the supply side, A, *Fig. 9*, need not affect the discharge, if such variation is also caused to occur on the efflux side B.

The following method of automatically recording the "head," or difference of water-level on the supply and efflux sides of such an orifice, was devised in connection with the experiments last referred to. The apparatus consists of two floats, immersed in the water at the upper and lower sides respectively of the orifice. The relative motion of the floats is reduced to one-third for a differential

G

recording-gauge which is of the revolving disk form.   The
recording-pencil is actuated solely by the relative water-
levels on the two sides of the
orifice, without reference to the
actual water-levels.   The arrangement of pulleys to effect this is
shown in *Fig. 10.*   The diameters
of the double pulley A, which
turns on a fixed axle, are 4 to 1 ;
those of the double pulley B,
which turns on an axle suspended
from A, are 3 to 1.   C and D are
fixed guiding-pulleys.   W is a
weight hung on the end of the
continuous cord which passes from
the upper float round C and A.*

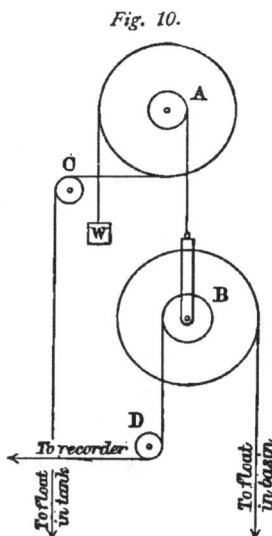

Fig. 10.

82. With small circular orifices,
having a free discharge, Mr. J. G.
Mair-Rumley has ascertained that
with sharp and square edges, and
for diameters varying between 1 inch and 3 inches, the coefficient of discharge equals 0·61 within 1 per cent., for
heads from 9 inches to 24 inches, at temperatures between
51° F. and 55° F.   To show the effect of the temperature
upon the discharge, the experiments summarised in the
following table were also made by him : †

SHARP-EDGED ORIFICE, 2½ INCHES IN DIAMETER.

Head.	Temperature.	Coefficient.
21 inches	57° F.	0 604
,,	92° F.	0·604
,,	110° F.	0·604
,,	153·5° F.	0·607
,,	179° F.	0·607

* Minutes of Proceedings Inst. C.E., vol. c. p. 281.
† *Op. cit.,* vol. lxxxiv. p. 427.

These results show that a range of temperature of 122° F. only increases the discharge by ½ per cent.

Again, from a series of experiments by Prof. W. C. Unwin on thin-edged orifices, between the temperatures 61° F. and 205° F., the conclusion is drawn that this range of temperature has extremely little influence on the discharge.*

83. An accurate method of observing the height of the surface of water for all gauging purposes is afforded by the "hook-gauge." This apparatus consists of a rod, screwed at one end, and at the other pointed and turned up to form a hook. The hook is attached to a frame, and is capable of being moved longitudinally by means of a nut. A scale is fixed on the frame, and over it is a vernier connected with the screw, enabling the level of the water to be read off to $\frac{1}{1000}$ foot. The frame is fixed vertically, the hook being allowed to dip beneath the surface of the water. When it is desired to ascertain the level of the surface of the latter, the point of the hook is raised by means of the nut until it touches the water-surface from below, an operation which can be effected and observed with great nicety. The level of the sill of the gauge can be read on the scale by a straight-edge levelled upon the sill and the point of the hook.

The gauge is situated some distance away from the notch, so as to measure the level of the surface of the still pond behind it, and must be sheltered from the wind and its effects.

84. One of the most ancient applications of the science of measuring water in motion relates to gauging for irrigation purposes. The system employed for this object by the Moors before their expulsion from Spain, and still used there, affords interesting evidence of views held eight

---

* *Philosophical Magazine*, Fifth Series, vol. vi. p. 281.

centuries ago upon the subject of the division of a stream of water into fractional parts. By the plan referred to, the water is conducted along a level masonry channel, at a low velocity, and is caused to flow over a sill or step. At a distance of about 3 feet beyond this point it falls over another sill. In the intermediate space is fixed a small pier, which divides the breadth of the channel into two portions, the ratio of which is nearly proportional to the two subdivisions into which the water is to be separated, *Fig. 11.* The up-stream point of the pier consists of a cut-water movable horizontally, the point of which, when in a line with the pier, almost touches the first sill. By moving the cutwater, the stream of water flowing over the sill may be divided approximately in the same ratio as the cutwater divides the breadth of the channel.

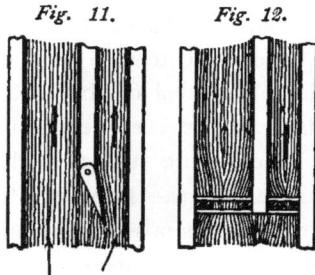

Fig. 11.   Fig. 12.

By another plan, instead of a movable cutwater, each portion of the channel is arranged so as to be closed if desired by a number of vertical bars of wood, which fit in grooves formed in the sides and bottom of the channel and are held at the top by bars placed across it, *Fig. 12.* The division of the water between the two portions of the channel is effected in the desired ratio, by taking out of each of them a number of bars corresponding with the quantity of water to be passed down that portion.

The principal characteristics of a good "module," as a gauging apparatus for irrigation purposes is termed, are simplicity of the several parts, freedom from friction or causes of derangement, constant discharge under varying heads, exact measurement, and facility of inspection by all interested parties.

85. The "Milanese" module is one of the best known. Its leading principle is the discharge of water through an orifice under constant pressure. In this connection the established unit, termed the "oncia magistrale," is the quantity of water which flows freely into the atmosphere under the action of gravity through a rectangular orifice 4·11 inches long by 7·86 inches high, under a head of 3·95 inches measured above the upper edge of the orifice. The discharge is about 0·87 cubic foot per second. The water is admitted from the main canal, that supplies it into a chamber, which, according to law, should be 20 feet long, and is introduced to steady the flow of the water. The floor is sloped towards the orifice, which is formed in a slab of stone, the actual edge being frequently made of metal. It is the duty of a public official to prevent the water from rising above the regulation height over the orifice. Any multiple of the "oncia" is assumed to be obtained by proportionately increasing the length of the orifice, and this naturally leads to inaccuracy in favour of large consumers of water.

86. In the ingenious module devised by Juan de Ribera, water is admitted through a screen into a chamber on one side of the supply-canal. In the floor of the chamber is fixed a wrought-iron plate, pierced by a circular hole, in which a plug is suspended from a float, *Fig. 13.* As the water in the chamber rises or falls, the annular space through which the water passes round the plug is, owing to the shape of the latter, decreased or increased. The plug must be accurately formed in order to secure constant discharge with a varying head of water.

Fig. 13.

If R is the radius of the circular hole, r the radius of the plug at any point, when h is the height of water above the plate and Q is the discharge,

$$Q = c\pi(R^2 - r^2)\sqrt{2gh},$$

c being an experimentally determined coefficient ;

therefore
$$r = \sqrt{R^2 - \frac{Q}{c\pi\sqrt{2gh}}}.$$

If $c = 0.61$,
$$r = \sqrt{R^2 - \frac{Q}{15.4\sqrt{h}}}.$$

The most serious drawbacks to the employment of this form of module lie in the loss of head entailed by it, and in the disturbance due to the deposition of mud on the floor of the chamber. By a modification of the arrangement, the plug is fixed so as to work horizontally, and the loss of head is thereby considerably reduced.

On many irrigation works the water is measured at its various discharging points by causing it to flow over notch-gauges—a method that causes but slight loss of head and ensures most accurate measurement. The water is admitted by sluices into chambers, out of which it passes through slits or grids, in order to steady its motion, into a second chamber, at the end of which the gauges are fixed.

87. When the flood-waters of streams are stored in reservoirs for purposes of supply, the fixed quantity of compensation-water sent down the stream (§ 34) is generally measured through orifices. A usual practice is to admit the water through sluices into a chamber, whence it is discharged through orifices at the opposite end. In order that those interested, who may not be skilled in hydraulics, may at any time convince themselves that the quantity agreed upon is passing through the compensation

orifices, a gauging-chamber of rectangular form is often provided below the level of the orifices. Into this chamber there is an opening situated immediately below the orifices, of such breadth that the jet of water issuing therefrom, under the head which gives the required discharge, leaps clear across it. To find this breadth, b, assume that the water issuing from the bottom of the orifices, describes the parabola due to its mean velocity.

This velocity is

$$c\frac{2}{3}\sqrt{2g}\,\frac{h_2^{\frac{3}{2}} - h_1^{\frac{3}{2}}}{h_2 - h_1};$$

h_2 being the height from the water-surface to the lower edge of the orifice ;

h_1 being the height from the water-surface to the upper edge of the orifice ;

and c being an experimentally determined coefficient.

If d is the depth of the opening below the lower edge of the orifice, and t is the time occupied by the water in falling through that depth,

$$d = \frac{1}{2}g\,t^2,$$

therefore

$$b = c\frac{2}{3}\sqrt{2g}\,\frac{h_2^{\frac{3}{2}} - h_1^{\frac{3}{2}}}{h_2 - h_1} \cdot t$$

$$= \frac{4}{3}c\,\frac{h_2^{\frac{3}{2}} - h_1^{\frac{3}{2}}}{h_2 - h_1} \cdot \sqrt{d}.$$

By means of a shutter, the jet of water issuing from the orifice may be diverted instantaneously into the chamber, when it is desired to measure the quantity passing through the orifice ; and after allowing the water to flow into the chamber for a given time, it may be instantaneously restored to its former course. Thus the quantity discharged by the orifice in a given time can be directly measured with great accuracy.

88. It is a common practice to estimate the rate of pumping water by multiplying the capacity of the pump-barrel, by the number of strokes per unit of time, and then to make a deduction of 2 to 5 per cent., assessed according to individual judgment, from the quantity.   This method can only be regarded as affording a rough approximation to the truth. With a long suction-pipe and a short delivery, it may happen that a greater quantity of water passes through the pump at each stroke than is due to its capacity— owing to the momentum of the water in the suction-pipe causing both valves to be open together momentarily. With a long delivery it usually happens that some water passes back through the valves before they close ; and consequently the delivery at each stroke is less than the capacity of the pump-barrel, by an amount that varies in different cases, and is termed the slip.

89. We proceed to consider certain methods of gauging flows of large amount, when permanent apparatus of the kind hitherto described is inapplicable.   The principle involved is that of ascertaining the average velocity of the stream from a measurement of that of individual stream-lines, and establishing a relation between this velocity and the variable level of the surface of the stream.

The most accurate measurement according to this process is obtained by observation of the velocity of the flow by current-meters.   A narrow stream may be spanned by a bridge from which current-meter observations are conveniently made.   In this case, the meter is fixed upon a staff that is moved vertically and horizontally along the bridge, and so attached to the latter as to maintain the current-meter parallel to the axis of the stream at every point of observation.   In a broader stream, or when traffic must be provided for, high poles may be fixed on either bank, and a strong wire attached to the top of each tightly stretched across the river. A small carriage travels on this wire, and can be moved to any position upon it, by means of a continuous cord attached to the carriage-and passing over pulleys on the poles. By winding this cord on drums of known diameters, the distance of the carriage from each bank of the river can easily be arrived at. A second cord passes over a pulley on one of the poles and over another one on the travelling carriage, and to it is attached the frame which carries the current-meter, so designed as to maintain the axis of the meter parallel to the line of the current. By means of a guiding-wire controlling the frame and steadied by a heavy anchor, the current-meter can be maintained in any desired vertical. If there are circumstances to prevent the adoption of either of these expedients, the frame to which the current-meter is attached may be fixed in front of a boat, or between two boats ; and its position, for any observation, may be fixed by observations with a theodolite on the bank, or measured by chains or wires. The section of the stream at the point in question, which should be in a straight reach, must be accurately ascertained by sounding or by levelling. In the first case mentioned this is readily done from the bridge ; in the second case, by a sounding-wire passing over one of the pulleys of the travelling carriage ; and in the third case, by sounding from a boat—the position of the latter at each sounding being accurately ascertained by ranging the boat in the line of section from the bank and fixing its position by angular observations.

90. A current-meter consists essentially of a revolving fan shaped like a ship's propeller, about six inches in diameter, furnished with means of counting the number of its revolutions when free to revolve in the current. Current-meters may be divided into two classes : (1) those which contain mechanism for counting the revolutions of

the fan and have to be taken out of the water for each reading ; and (2) those which indicate at the surface, by electrical contact, each revolution or a given number of revolutions. In the latter kind, the time between certain of the signals is observed, and the rate of revolution of the fan of the meter is thus arrived at.

Of class 1, the Revy, Woltmann and Moore meters are examples. In these the fan is provided with a mechanical counter which indicates the number of revolutions made by it, and with means of throwing the counter into and out of gear. A rudder is attached to the tail of the meter in order to keep the fan facing the current. Owing to the Revy and Woltmann meters being placed in position on rods, a considerable time is lost in lowering and raising them for reading, in order to take observations at different depths. This delay allows time for the level of the water to change ; consequently, the observations are not always taken under the same conditions. Owing to the method of suspension by a cord or chain, adopted in the Moore meter, it is much quicker in action. The chain is attached to a stirrup which carries bearings for a spindle whose line intersects the centre of gravity of the instrument, and is at right angles to it.

91. The Harlacher and Deacon current-meters are examples of class 2. From a water-tight case a spindle, carrying the fan at its extremity, projects through a well-made bush. A worm on the spindle, inside the case, turns a small spur-wheel ; and at every revolution contact is made between the latter and an insulated conducting strip to which an electrical wire is attached. Another wire is attached to the case of the instrument, and both are in circuit with a primary battery and an electrical indicator at the observing-station. The indicator signals each time that contact is made. This may be caused to occur at

any number of revolutions of the fan, by suitable arrangement of the pitch of the worm and the number of teeth in the spur-wheel. By observing the time that elapses between successive signals, the rate of revolution of the fan is ascertained. The case is filled with paraffin or other mineral oil, to ensure constant friction and to improve the electrical insulation. Before use, the meter is accurately rated by moving it through still water at known velocities, the rate of its motion relatively to the water being ascertained for a number of different speeds ; from which data a curve is plotted, giving the velocity of the current for all rates of revolution of the fan within the range of the observations. Many of the formulæ proposed for the calculation of these data give results near the truth ; but the most satisfactory method of rating current-meters for practical work, is that described.

All the parts of current-meters should be made of non-corrosive metal, so that the friction may not vary, by the rusting of the parts.

92. In the current-meter invented by Prof. A. R. Harlacher,* instead of signalling by an electro-magnet, an electrically controlled counter may be put in circuit with the wires from the meter. This meter is constructed so that the mean velocity in a vertical section of the stream may be obtained at one observation. With this object, the small water-tight horizontal cylinder from which the shaft of the fan projects, is screwed to a sleeve that slides on a hollow cast-iron rod, being counter-balanced by a rudder to preserve the direction of the meter. To reduce friction in sliding, friction wheels are fixed within the sleeve bearing against the rod. The latter has a solid point at

---

* Harlacher, 'Die Messungen in der Elbe und Donau und die Hydrometrischen Apparate und Methoden des Verfassers'—Leipsic, 1881.

the bottom and is slotted up one side ; it is firmly planted in the bed of the river, in the vertical section of which the mean velocity is required.   An arm from the meter passes through the slot to the inside of the rod, and to a point on this arm a woven copper wire is attached, by which the meter may be raised or lowered.   The drum on which the wire is wound is attached to the vertical rod.   The axis of this drum is geared to a train of clockwork, the movement of which is controlled by a fan.   The meter is allowed to fall from a short distance above the surface of the water, and the counter is read and the time noted at the instant it touches the surface.   The meter descends uniformly down the rod, and the time at which it reaches the bottom of the stream is also noted and the counter is again read. From the data thus obtained, the mean velocity in that vertical section is ascertained by repeated trials.   For practical use, the battery, the electrical indicator and the rod carrying the meter are placed on a float, which in large rivers is anchored, and in small rivers is fastened by gye-ropes and poles.   By slackening the rope on one side and tightening on the other, the float is brought into a new position.   The rod is then ranged in line by a person on the bank, and observations in successive verticals may be thus taken.

93. In a modification of the Harlacher current-meter, a worm on the screw-spindle gears into a small toothed wheel, which makes one revolution for every hundred revolutions of the screw.   At every revolution of this wheel, electrical contact is made, by which a signal is given to an observer by the exposure of a coloured disc.

The Deacon current-meter is practically of this form. The electro-magnet rings a bell at the observing station, and may be geared so as to ring at every revolution or at every twenty revolutions, as desired.   This meter is not designed to give the mean velocity in a vertical at one operation.

94. The proximity of the verticals in the cross-section of a stream at which observations should be taken, depends on the limit of accuracy desired. For a stream, say, 200 feet wide, observations might well be taken every 10 feet apart, horizontally, and at every 1 or 2 feet vertically from the surface, according to the depth—unless the mean velocity in each vertical is measured at one operation. The fan of the meter must always be submerged when taking an observation. During the progress of current-meter observations, the level of the surface of the water is frequently read on graduated staffs placed at both banks of the stream.

The mean velocity in each vertical line is arrived at, either directly or by plotting the vertical curves at each position and finding their area, if separate observations are taken at different depths. The mean velocity in the whole cross-section is found by plotting the mean vertical velocity at each point of observation, finding the area of the curve thus formed and dividing it by the breadth of the stream. This mean velocity is referred to the water-level corresponding with the time of the observations. The areas of the curves may be accurately found by the planimeter, or may be calculated by the aid of Simpson's rule.

A series of observations being made at a certain section of the stream, corresponding with surface-levels from the lowest dry-weather flow to the highest flood, a curve may be plotted giving the rate of discharge at the various levels assumed by the surface of the river. By means of the record afforded by an automatic float-gauge, registering continuously the varying level of the river, a complete statement of the flow may be derived from the curve of discharge thus constructed.

The variation of velocity measured at points in any vertical, may be more than 25 per cent. from the average velocity in the course of a few seconds. Hence the necessity for allowing current meters to run for some considerable time during the course of every observation, so as to obtain the average motion.

95. It frequently happens that, owing to the presence of silt or floating weeds, or to other considerations, current-meters cannot be advantageously employed for gauging the flow of streams. The most serviceable method in such cases is to observe the run of submerged or nearly submerged floats over a measured course of the stream unaffected by wind. Such floats may be disks, as " surface-floats " ; or rods weighted so as to float vertically in the water ; or a combination of the surface-float with a sub-surface ball attached to the former by a cord or light chain. The results given by the use of a double float are vitiated chiefly by three causes : (1) The surface-water generally moves either at a greater or less rate than the sub-surface weight ; consequently the course of the latter is situated at a less depth than the length of the connecting cord would indicate. (2) The same reason causes the sub-surface float to travel at either a greater or less rate than the water surrounding it. (3) The velocity of the lower float is affected by the action of the current on the connecting cord. It results that the efficiency of sub-surface floats decreases with the depth, and the mean velocity past a vertical, as indicated by it, is greater or less than its true value, according as the surface velocity is greater or less than the mean.

Instead of the surface-float, a second ball, weighted to float just below the surface, may be employed. The velocity v measured by this twin float, is the mean of the surface-velocity v_s and the velocity v_d at the depth of the sub-surface float. As the two former can be observed, the latter is obtained from the equation

$$v_d = 2\,v - v_s.$$

For surface-floats, disks of wood loaded to float almost submerged, provided with a stem to mark their position, may be employed. But of all kinds in use, the " rod-float " is the most effective and reliable. The form of this float extensively used by us, consists of a light wooden rod, weighted at one end by a small lead ring (cut off a piece of gas-pipe) and provided at the top with a cork float, through which it is thrust a few inches. These floats are simple and inexpensive, and if care be taken that they reach nearly to the bottom of the stream, and acquire a steady velocity before their actual " course " is observed, a close approximation to the mean vertical velocity may be obtained by their use. Colonel A. Cunningham has estimated that a rod-float 0·94 of the depth of the stream gives the true mean velocity past a given vertical ; and, making allowances for the inaccuracies of double floats, his experiments on the Ganges Canal confirm this view.* The rod-float presents a ready means of arriving at the average rate of flow of a stream, by numerous observations at points close together in the cross-section.

96. When a stream is not too deep, and its velocity is not very small, Darcy's modification of the Pitot tube is sometimes employed for its measurement. The original tube of Pitot consisted simply of a vertical glass tube with a right-angled bend at the bottom presenting the mouth of the tube normally to the direction of flow. By making the mouth small, the velocity of flow is not interfered with ; and the height h, to which the water rises

* ' Roorkee Hydraulic Experiments,' p. 246—Roorkee, 1881.

in the tube, is that due to the impact of the water moving
with the velocity v of the stream.

That is, $$h = \frac{v^2}{2\,g}.$$

The chief difficulty affecting the use of Pitot's tube in
its simplest form, arises from the inconvenience of reading
it near to the surface of the water. This objection is
removed in the Darcy instrument, which consists essentially
of two Pitot tubes, fixed in planes at right-angles to one
another, united at the top, and connected with a small
flexible pipe, provided with a cock. The apparatus is
clamped to a vertical rod, and is placed in the stream so
that the mouth of one of the tubes faces the current, and
that of the other is at right-angles to it. The difference in
height of the water in the two tubes is therefore a measure
of the velocity of the stream at that point. To facilitate
the reading of this difference, a little air is exhausted from
the tubes by the flexible pipe, and the cock is closed. This
causes the level of the water in both tubes to rise propor-
tionally to the reduced pressure, but the difference of level
remains constant. Owing to the fluctuation in the velocity
at all parts of streams, several measurements must always
be made with the instrument at each point of observation—
the average of these being taken as the true mean velocity.
This instrument, however, is not applicable to very small
velocities of flow, nor can it be used in deep channels.

97. The hydrodynameter introduced by M. Perrodil
indicates the velocity of a current by measuring the torsion
of a wire, produced by the impact of the stream against a
submerged vane.* The instrument, like the Darcy tube,
does not require a time observation, and it need only be read
once at each point—the average torsion of the wire being of

* Annales des Ponts et Chaussées, 1877, p. 467, and 1880, p. 11.

course taken ; but its use is fraught with some practical disadvantages which preclude its general employment.

98. Observations have shown that, in any vertical line in a stream, the maximum velocity may be situated at the surface, or, as is generally the case, somewhat below that level, down to $\frac{3}{10}$ths of the depth ; but the mean velocity is with little variation situated at $\frac{2}{3}$ths of the depth.*

The curve of velocity in a vertical may be approximately represented by a parabola, of which the axis is situated horizontally at the position of maximum velocity. The curve, however, falls actually inside the parabolic curve near the surface and near the bed of the stream. The effect of wind blowing down stream is to raise the position of maximum velocity, and *vice versa.*

In any cross-section, the mean velocity in a vertical plane generally attains a maximum value near to the middle ; but its exact position is governed by the shape of the stream and the contour of its bed at and near to that section. The variation in the magnitude of the surface-velocity in a cross-section follows, to some extent, that of the mean velocity. The fact of the surface-velocity in a stream being generally less than the maximum velocity, has been explained by the late Dr. James Thomson to be due partly to the friction of the atmosphere, but especially to the eddying motion of the water—that from the bottom being constantly moved towards the surface, where its upward motion is necessarily stopped.† It consequently tends to remain there longer than at other points in the vertical.

99. In finding roughly the discharge of streams, the three following propositions, established from a large number of experiments on various rivers and streams, may be usefully noticed.

* Wagner, ' Hydrologische Untersuchungen,' p. 37—Brunswick, 1881.
† Proc. Roy. Soc., vol. xxviii. p. 114.

(1) To determine the mean velocity in a cross-section, from observation of the maximum surface-velocity—Prof. von Wagner has proposed, as the result of experiments on twenty-four large and small streams, the formula[*]

$$v = 0\cdot705\ v_c + 0\cdot003\ v_c^2\ ;$$

where v_c is the maximum surface-velocity in feet per second, and v mean velocity of the entire cross-section.

(2) To find the mean velocity from observation of mean velocities in equidistant verticals—since the mean velocity in every vertical line is found approximately at $\frac{2}{3}$ths of the depth, the velocity at this depth may be measured by one of the methods described, at a number of equidistant points across the stream ; and the mean velocity of the entire cross-section may be deduced from them.

(3) To determine the mean velocity from an observation of two velocities in each one of a number of equidistant verticals,—

Colonel A. Cunningham found, from the experiments previously alluded to, the mean velocity in a vertical line to be given by the formula [†]

$$u = \frac{I}{2}\left(v_{0\cdot211\,H} + v_{0\cdot789\,H}\right)\ ;$$

where u is the mean velocity in a vertical, and $v_{0\cdot211\,H}$ and $v_{0\cdot789\,H}$ are the velocities at $0\cdot211$ and $0\cdot789$ of the depth below the surface. By means of double floats at these two points, attached to a surface-float, the mean of these two velocities may be found by a single observation.

100. The flow of water through pipes, whether they be mains for town supply, or distributing- or service-pipes supplying districts or individual consumers, is gauged by

[*] *Deutsche Bauzeitung*, No. 82, 1882, p. 480.
[†] Minutes of Proceedings Inst. C.E., vol. lxxi. p. 18.

being caused to pass through some form of water-meter. These instruments may be classed as " positive " and " inferential " meters.

The former measure by displacement, the water being caused to flow into and out of chambers of known capacity ; and the number of times that each chamber is filled is mechanically counted and recorded by automatic gear. In the latter class, the water impinges against vanes or disks, and the velocity of its motion is indirectly measured either by its effect in turning a small turbine or water-wheel, the revolutions of which are mechanically counted, or by its momentum being used to compress a spring, or to move a balanced disk into such a varying position that the pressure against it is constant whatever be the velocity of the impinging water. In the last mentioned case, the position of the disk at any moment approximately indicates the velocity of the water.

101. The essential characteristics of a perfect water-meter are : (1) accurate registration of the quantity of water passing through it, whether large or small; (2) ability to perform its work without causing a material loss of head in the supply-pipe ; (3) cheapness and simplicity ; (4) ease of attachment and repair ; (5) freedom from excessive wear of the working parts.

The entire apparatus must be made of such materials that corrosion will not interfere with its working. It must not give rise to concussion in the supply pipe, must never stop the supply, and should occupy but little space. It is difficult to obtain a meter that complies with all these conditions, but it generally happens that some of them are, in particular cases, of less importance than the others ; and a meter may generally be chosen suitable to the circumstances under which it is to be employed. Positive meters may be reciprocating in action or they may be rotary.

102. Among the first mentioned type, the Duncan, Kennedy, Frost, Schönheyder and Frager meters are often employed for high-pressure supplies ; and the Parkinson and Tylor meters for low-pressure supplies. The Kent meter is a good example of the second type.

The Duncan meter is practically a double-cylinder water-pressure engine having D slide-valves faced with lignum vitæ. The pistons are provided with cup-leathers. This apparatus may be used as a motor, in which case, of course, part of the available head of water is used in the work.

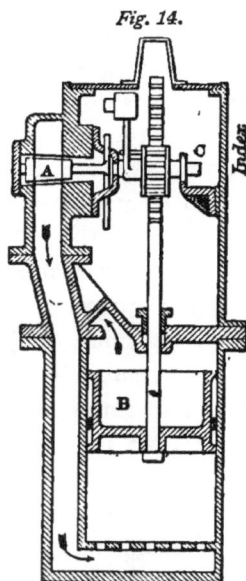

Fig. 14.

¾-INCH KENNEDY METER.
A, four-way cock ;
B, piston ;
C, counting-gear.
Scale 1/10.

103. The Kennedy meter, *Fig. 14*, consists of a cylinder provided with large ports at the top and bottom. Within it works a long vulcanite piston, which fits the cylinder at its upper and lower extremities. In the middle space, a ring of indiarubber rolls up and down upon the piston as it moves, making a water-tight and nearly frictionless joint with the cylinder. The piston-rod passes through a stuffing-box in the cylinder cover, and is attached to a rack which works a pinion on a horizontal shaft. In line with the shaft is situated a four-way plug-cock, connecting the inlet and outlet alternately with the two ports of the cylinder, and it is fitted with a two-armed lever. A tumbling-weight which works loosely on the shaft is lifted by small arms on the pinion, so as to overbalance and fall on the arms of the lever alternately,

as the piston reaches the upper and lower ends of its stroke, thus reversing the connection of the inlet and outlet with the two ports. ' The tumbling-weight, after striking the lever, falls on a buffer which gradually arrests its motion.

The machine requires occasional attention to prevent the moving parts from sticking, and to keep the plug-cock in action ; and, in common with other reciprocating direct-acting single-cylinder meters, a slight shock in the pipes is caused by it at each reversal of the plug-cock. One great advantage possessed by this meter, is that the recording gear worked from the shaft measures the length of the stroke of the piston, and thus also the actual volume of water displaced by it. By means of suitable gearing the apparatus records continuously, notwith-standing its reciprocating motion. The indiarubber ring is found to last longer if the cylinder is lined with brass.

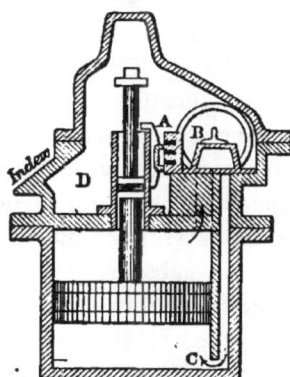

Fig. 15.

I-INCH FROST METER.

A, auxiliary slide-valve ;
B, main slide-valve ;
C, main cylinder ;
D, counting-gear.
Scale $\frac{1}{10}$.

104. In the Frost meter, *Fig. 15*, a piston, provided with leathers, works in a brass-lined cylinder, the upper and lower ends of which communicate with ports, on each side of a main outlet - port. A horizontal D slide-valve, connecting in turn each port with the outlet, is actuated by two small pistons working in cylinders situated one on each side of the valve. An auxiliary vertical slide-valve, actuated by tappets on the main piston-rod, admits water alternately behind one and the other of the small pistons that drive the main horizontal valve, pushing it

over at each end of the stroke of the main piston ready for the next stroke. The counting-gear is worked from the reciprocating piston by the usual train of mechanism.

105. The Schönheyder meter consists of a circular cast-iron case furnished with inlet- and outlet-pipes, and contains three horizontal cylinders set at angles of 120° with one another. The three pistons working in these cylinders form one piece with a central valve which works on a horizontal valve-face containing the inlet- and outlet-ports. In the centre of this valve is a circular recess which fits over a central vertical roller of a diameter smaller than its own, and the spindle of which is attached to the case. Each point of the pistons has thus a circular motion, to permit which the cylinders have a transverse sliding-motion controlled by rollers. The ebonite valve-face contains three openings communicating directly with the interior of the meter-case, arranged symmetrically about a central outlet-port which communicates with the outlet-pipe. In the valve are three openings, also symmetrically arranged, through which the water passes from the ports in the valve-face to the cylinders and is subsequently exhausted to the outlet-port. The pistons are furnished with cup-leathers. The counting-gear is worked off the circular motion of the pistons.

106. The Frager meter has two vertical cylinders placed side by side. In these cylinders work chambered pistons with loose rods passing through them. The loose piston-rods, to the tops of which D slide-valves are attached, only move as far as is necessary for them to actuate the valves. Their travel is effected by the upper and lower ends of the piston-chambers alternately striking bosses at their lower extremities. When either piston has travelled nearly to the end of its up-stroke, the back of the chamber comes into contact with the boss on the piston-rod and forces the valve forward. Similarly, at the end of the down

stroke, the front of the chamber in the piston strikes the
boss on the piston-rod and withdraws the valve. There are
three ports under each slide-valve. The upper and lower
ports in the valve-face above one cylinder communicate
with the ports of the other cylinder, and the pistons
reciprocate alternately. The central port leads to the
outlet. Consequently, one end of each cylinder is always
open to the supply and the other to the outlet. A mechanical counter records the delivery in terms of the number
of strokes made by the pistons.

107. For low-pressure supplies, the Parkinson and Tylor
meters are much used. The former works on the principle
of the Parkinson wet gas-meter, and is found satisfactory
in practice.

The Tylor "bascule" meter consists of a tumbling-
vessel, divided into two compartments of known capacity,
rocking on a horizontal axis which may be placed over a
house-cistern. The supply-pipe delivers into the two compartments alternately. When one is full, the vessel tilts
over, emptying its contents into the cistern; and the other
compartment is simultaneously brought under the supply-
pipe and begins to fill. Automatic gear of the ordinary
ratchet-and-pawl type, counts and registers the number of
times each compartment is emptied. A ball-valve cuts off
the supply of water when the cistern is filled,-and puts the
meter out of action. Large quantities of water may be
thus measured by being passed through a number of similar
apertures, the amount passing through one of them being
measured directly by the meter, and the quantity thus
registered being multiplied by the number of the orifices
employed.

108. The Kent meter, *Fig. 16*, has a cylindrical working-
chamber tightly fitting into a cast-iron case through which
the inlet- and outlet-pipes pass on opposite sides of the

working-chamber. This latter is divided by a straight vertical diaphragm into two unequal segments, in the larger of which a rotating piston works. The smaller segment is

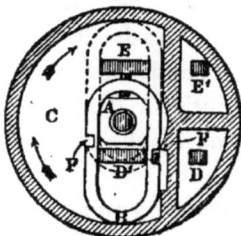

Fig. 16.

KENT METER.

A, square axle;
B, piston;
C, working-chamber;
D D, inlet-ports; E E, outlet ports.

subdivided into two equal parts, into one of which the water is admitted and into the other the water is delivered after passing through ·the meter. The object of these chambers is to allow the water to enter and leave by ports on either side of the piston. In the larger segment, at the centre of the circle, a pin is fixed, upon which a square axle turns. The piston, which has straight parallel sides, but is rounded at the ends, is hollowed inside to fit

accurately over the square axle, on which it slides. It is of such a length that when close to the axle at one end, it touches the side of the chamber at the other; and when parallel to the diaphragm it fits closely to it. In this position, and touching the sides of the chamber at one end of the diaphragm, the port from the inlet-chamber communicates with the space within the piston, and when slid along the diaphragm to the other end the outlet-port is in communication with it.

The action is as follows :—Starting with the piston lying close to the diaphragm at the inlet side, water enters between it and the diaphragm and causes the piston to rotate through 180° with the square axle, which brings it over the exhaust port. The water then enters the piston through the inlet-ports between the axle and the circular end lying close to it, and causes the piston to slide parallel to the diaphragm, until it reaches the other end of its stroke, which is the position started from. A lever from the axle works the counting apparatus. The piston is of vulcanite and the working parts are of delta metal. On the inlet side a screen is fixed to catch grit, &c., as in other positive meters.

Where a supply of water is laid on through a cistern, and the latter is nearly full, a mere dribble of water passes through the pipes ; a "positive" meter is then the only form which gives correct registration of the quantity of water supplied.

109. Of the "inferential" type, the Siemens, Tylor, Sporton, Deacon and Venturi water-meters may be referred to.

There are two forms of the Siemens inferential meter,—the English and the German. In the former, the water enters a hollow rotating wheel through a central funnel, issuing by tangential passages into the casing surrounding it, as in a Barker's mill. To prevent excessive speed of rotation, vertical blades are attached to the periphery of the wheel and act as a brake by the resistance of the water—causing an approximately constant ratio to obtain between the velocity of the flow through the meter and the rate at which the wheel revolves. This form of meter reduces the pressure considerably, and the orifices are apt to become choked with solid matter, which interferes with the accuracy of registration.

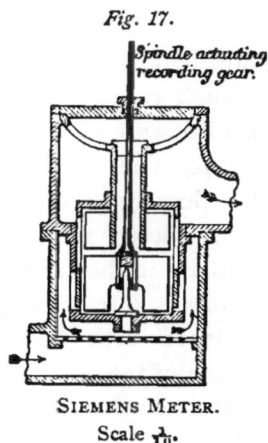

Fig. 17.

Spindle actuating recording gear.

SIEMENS METER.
Scale $\frac{1}{10}$.

In the German form of the instrument, *Fig. 17*, these objections are obviated. It consists of a gun-metal, brass, or in larger sizes, cast-iron cylindrical chamber, in which revolves a fan, consisting of a vertical spindle with four plane vanes symmetrically fixed at right-angles to it. These vanes almost brush the sides of the chamber. The water from the supply-pipe enters tangentially at the periphery of the fan and impinges normally against the vanes, setting them in motion and revolving with them in the chamber. The whirling water next passes into an upper part of the chamber above the vanes, where its rotation is checked by a diaphragm ; it then flows away through the outlet. The number of rotations of the fan is registered by recording gear actuated by the spindle at the top of the meter, and affords a measure of the water passing through it. This meter gives correct results except for small driblets, such as occur when supply-pipes deliver into cisterns through slowly-acting ball-valves.

In order to prevent small quantities of water from passing through a large meter unregistered, a by-pass pipe is sometimes placed round the main inlet and outlet, and is provided with a meter of smaller size. When the velocity falls below a certain point, a weighted throttle-valve closes the main inlet, and the water is diverted through the auxiliary meter, which is more sensitive to small flows.

110. The Tylor and Sporton meters closely resemble the German form of the Siemens instrument. In the Tylor meter, the water enters from below almost tangentially to the vanes of the fans, and so causes the latter to rotate. It then passes upwards between radial ribs furnished with horizontal projections, which deflect a portion of the flow back to the fan. This action is greater as the velocity of flow increases, so rendering the meter accurate through a considerable range of working. Vertical corrugations are formed on the case at intervals round its circumference, acting as "vortex chambers" and retarding the speed of the fan as the flow increases. The fan is made of phosphor-bronze, and the case of gun-metal.

This meter has been made self-registering, by recording on a strip of paper moved forward by clockwork each 500 gallons that pass through the meter.

In the Sporton meter, the water is admitted below the fan and caused to support its weight, thus rendering the friction of the pivot insignificant. This, like the Tylor meter, has been made to record automatically its measurement. To effect this, a drum hung over one of the index-spindles of the recording-gear has a vertical reciprocating motion imparted to it by clockwork, performing its stroke in 24 hours. The index-spindle carries an arm on its upper extremity, which is bent over so as to approach the outside of the drum, upon the paper covering of which it traces a diagram as it is carried round by the rotating spindle.

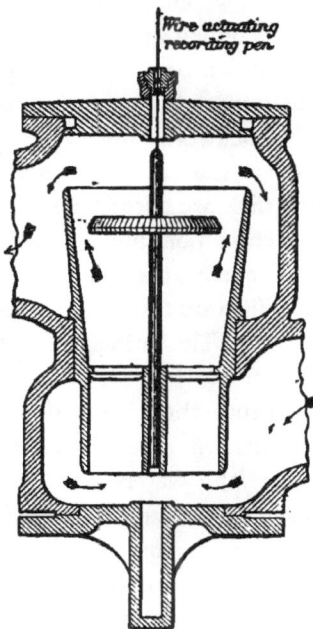

Fig. 18.

Deacon Meter.
Scale $\frac{1}{10}$.

111. The Deacon differentiating water-meter is suitable for measuring either large flows of water, or supplies to small districts, as applied in the waste - water - meter. Essentially, it comprises a disk suspended in an inverted frustum of a hollow cone, the lower end of which it accurately fits, *Fig. 18*. Attached to the lower end of the cone is a short cylinder containing four vanes, the boss of which is bored out to form a guide for the spindle of the disk. The object aimed at is to steady the motion of the

water by causing it to flow over the lip of the cylinder and to check any vortex motion before it enters the cone. The water enters from the bottom, flowing from the inlet-pipe round the cylinder, and leaves at the top where it overflows the lip formed by the upper edge of the cone, and so passes to the outlet.

The disk is weighted to the extent necessary to balance the pressure due to the velocity of the water past it, and it assumes such a position in the cone that, whatever the quantity passing may be, the pressure upon it due to the velocity of the flow is constant. The disk is supported by a wire which passes through the cover of the meter, and over a pulley with a small counter-weight to keep it taut, and actuates a pencil recording the heights of the disk continuously upon a diagram carried by a drum, which is rotated by clockwork, say, once in 24 hours. For each size of meter, the rates corresponding with different heights of the disk are ascertained by experiment; and the diagrams are correspondingly ruled to show the rate of flow through the meter at every point that can be assumed by the pencil in its travel.

112. The principle of the Venturi meter is worthy of attention.

From the observed reduction of pressure caused by a contraction in a pipe, the velocity and thence the rate at which the water is delivered through it is inferred as a question of hydrodynamics.

If v is the velocity, and p the pressure in the pipe and p' the pressure in the contracted portion; and if A is the area of the pipe, a that of the contracted portion, and w the weight of unit volume of water,

$$\frac{v^2}{2g} + \frac{p}{w} = \frac{\left(\frac{vA}{a}\right)^2}{2g} + \frac{p'}{w} \; ; \qquad (\S\ 65)$$

that is, $p - p' = \dfrac{w\,v^2}{2\,g}\left\{\left(\dfrac{A}{a}\right)^2 - 1\right\}.$

From this it follows that the velocity in the pipe is proportional to the square root of the reduction of pressure.

113. No assumptions have yet been forthcoming to form the basis of a satisfactory theory of any but the cases of steady motion already referred to (§ 65). In the channels of irregular form and surface which are met with in practice, the unsteadiness and internal disturbance of the flow become often so pronounced as to render direct gauging necessary to correct measurement.

The satisfactory application of gauging-processes is dependent in these cases upon the continuity of the observations, no less than upon the accuracy of individual measurements—the aim being to obtain *average* velocities or pressures.

Upon these considerations the suitability of instruments of all kinds for gauging purposes must be determined.

CHAPTER III.

THE COLLECTION OF WATER.

114. ALMOST all methods of collecting water for purposes of domestic or industrial supply can be summarised under two heads :—

(1) The water is obtained from the earth's surface, upon which it has fallen in the form of rain ; or (2) it may be derived from the sub-soil and rocks, in the course of its subsequent passage to the ocean.

115. (1) In many tropical, and in some European countries, rain is collected in tanks as it precipitates, or is led off the roofs of buildings into cisterns, often situated underground ; and the water thus preserved frequently forms the sole supply for dietetic purposes, as is the case in many parts of India. The Spanish Peninsula affords examples of the former class of tanks, employed for catching rain-water for irrigation purposes ; whilst in Venice may be found many underground cisterns, supplied from the roofs of houses, and in some instances provided with sand filters, through which the water passes to clear-water wells.

The famous tanks at Aden bear silent testimony to the inadequacy of any system of direct collection for ordinary water-supply (§ 270). The disadvantage of such sources is that, even in countries where the rainfall is heavy, the quantity available for the use of each individual is limited by the relatively small collecting-area. The quality of the water, however, where that which first falls is rejected and the cisterns are properly attended to, may be, under favourable circumstances, good ; because it is only subject to contamination from the gases and dust of the atmosphere.

116. The more generally practised way of collecting surface-water is to draw it from lakes or streams, when the quantity afforded by the latter is sufficiently constant ; or, if it is liable to considerable fluctuations, and is reduced at times below the supply required, to impound the flood-waters and so equalise the yield at the various seasons. The earth's crust, under certain conditions, itself forms the equalising reservoir—the water being drawn either from springs or shallow wells, or by means of intercepting tunnels or drains traversing water-bearing strata.

117. The considerations involved in the question of obtaining supplies of water from any sources whatever are : First, that it is sufficiently free from organic and mineral impurities; second, that the quantity available may be sufficient to satisfy the demands likely to be made upon it by the district that is to be supplied ; and third, that it can be conveyed from the source selected to the place of consumption at a reasonable cost.

Assuming a supply to be satisfactory, both as regards quality and quantity, and that physical circumstances point to the feasibility of using the source, the method of collecting the water in a convenient condition for its subsequent conveyance claims attention.

Where a river bank consists of sand or gravel, unjointed pipes, in the former case covered with gravel, may be laid in the bank, so that the water may percolate into them, and be thus partially filtered before its admission to the main collecting-tank or well. The supply of Oxford is drawn from the gravel beds through which the Thames flows, and a large district of London is at certain times similarly

supplied from Hampton. Lyons, again, derives a supply
from the Rhone, by shallow wells sunk in the gravel on the
bank of the river ; and this plan is extensively practised
both on the Continents of Europe and America.

118. When water is taken from a river or stream direct,
careful selection must be exercised in deciding upon a
suitable site for the intake-works. A position should be
chosen at which there is no tendency for the river to
deposit silt or other detritus ; consequently, either a concave
bank or a straight reach should be selected, and, if necessary,
rendered stable by suitable training-works. The river
Vistula was thus treated at Warsaw for a length of about
4½ miles, in order to render stable the concave shore for an
intake.*

The intake-chamber is generally situated above the
level of the river-bed, to prevent its being filled by the
material rolled along by the current ; but below the dry-
weather level of the surface of the river, so as to avoid the
ingress of floating matter. In countries where the whole
body of the water in a stream is liable to be cooled below
freezing-point, the latter precaution averts some of the
difficulties due to ground-ice. For, when water in such a
condition comes to rest in the intake-chamber, some of it
freezes ; but, by drawing the water from a sufficient depth
below the surface, it may be abstracted continuously underneath the floating ice so formed. The importance of
providing means to flush out the intake-chamber must not
be overlooked ; and the inlet should be protected by a
grating, to prevent the passage of any matter of sensible
size into the pipes.

119. To districts of supply which are situated sufficiently
below the intake-works, water is generally conveyed entirely

* 'Deutsche Vierteljahrsschrift für öffentliche Gesundheitspflege,' 1890,
p. 199.

by gravitation, unless high ground intervenes which cannot be tunnelled, and renders the resort to pumping necessary The latter is the only plan available where the difference of level is insufficient to produce the required flow under the action of gravity or where the district to be supplied lies above the intake; and, where practicable and a large supply is required, generally affords a cheap method of collection in first cost. Its adoption, however, generally implies that the dry-weather flow of the stream always exceeds the required supply, a condition of affairs which, unfortunately, seldom occurs when the water is of unimpeachable quality. The best surface-water supplies are usually obtained from streams of small size, draining comparatively limited catchment-areas from which sources of pollution can be easily excluded. In these cases, storage works (Chapter IV.) must be resorted to.

Stream-valleys affording suitable sites for impounding-reservoirs are, in this country, happily not uncommon. They should present an impervious stratum at no great depth below the bottom and sides, and should be free from faults, fissures or open joints—features that are often difficult of detection, even to a practised eye, and the discovery of which should form the object of a careful geological survey of the sites. Masonry walls are employed as dams where stone is plentiful, and of good quality. In narrow and rocky gorges such dams may be arched in plan—the sectional area being in that form much reduced; but, with foundations situated at considerable depths, or of a soft, yielding nature, requiring the pressure thereupon to be widely distributed, embankments of earth may be found the most suitable (§ 194).

120. To maintain as far as practicable the purity of water falling upon a catchment-area of light surface-soil, a system of land-drainage is useful, and reduces the loss from

I

evaporation. In impervious ground, unless it is very steep, open ditches are sometimes employed to intercept the water and convey it to the reservoir; and these, although less beneficial than covered drains, have a similar effect in reducing deterioration and evaporation of the rain-water.

121. It is sometimes practicable to allow only the clear water from the contributing streams, during their normal flow, to enter one of the series of reservoirs from which the supply is drawn. By an ingenious method, introduced by the late Mr. J. F. La Trobe Bateman, the separation of such water may be automatically effected. A shoot or flume is constructed to carry the stream operated upon almost, though not quite, across the channel which, immediately below it, is provided to carry the normal dry-weather flow of clear water to the reservoir (*Fig. 19*). In times of flood, the increased velocity of the stream causes it to leap completely across this channel, when it flows down a separate water-course to other reservoirs, in which it may be allowed to settle before being used.

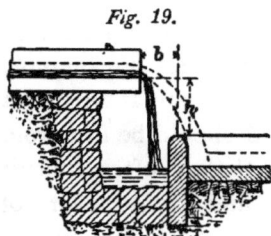

Fig. 19.

If v be the mean velocity of the stream in feet per second, when its water begins to be discoloured, h the depth in feet from the shoot to the edge of the clear-water channel, and d the depth in feet of the flow of the stream in the shoot,

$$v = \frac{2}{3} \sqrt{2gd},$$

and

$$h = \frac{1}{2} g t^2;$$

let b denote the horizontal distance, also in feet, from the end of the shoot over which the stream falls to the edge of the channel below, so that it may completely leap over this channel ;

then $$b = v\,t,$$

therefore $$b = \frac{2}{3}\sqrt{2\,g\,d}\;t,$$

$$= \frac{4}{3}\sqrt{d\,h}\;.$$

As the water, in falling, does not assume the curve due to the mean velocity of the stream, the value of b thus obtained is somewhat small, though not to such an extent as to matter in practice.

122. (2) Where the hydraulic surface over a tract of country nearly coincides with that of the earth, but no natural depressions exist in which the water may assume the form of lakes or streams, a supply may be obtained either from springs—if there are such ; or from shallow wells, intercepting-tunnels, or drains, by means of which the land-water is collected to centres from which it may be conveniently drawn. Many, probably most, towns have, at an early period of their history, been supplied from shallow wells, which have been abandoned as the increased pollution due to density of population and other circumstances has rendered this course obligatory.

The growth of London, and of many other towns, was at first confined to those districts adjacent to rivers, from the gravelly banks of which water could be readily obtained. In California there are upwards of 4000 shallow artesian wells, from 100 feet to 250 feet in depth, used for domestic supply and for irrigation. As, however, districts become more densely populated, or highly cultivated, such wells are very liable to contamination ; although, when sunk through impervious into water-bearing strata, the

I 2

dangerous surface-waters may be excluded by steining or
lining the wells with brickwork, masonry, or metal tubes,
down to the level of the impervious strata.

123. Deep-seated springs, being unaffected by surface-
pollution, generally form desirable sources of supply, both
for small districts and on the large scale. Examples of
small spring-water supplies are familar to every one, whilst
London (New River), Rome, Vienna and Frankfort may be
cited as instances of great cities thus supplied wholly or in
part. The supply of Lancaster is drawn from the Millstone Grit of the Wyresdale Fells through a system of
stoneware pipes carried into the springs, and so efficiently
contrived that the water is not exposed to light until it
reaches the consumers.*

124. Instead of numerous shallow wells, intercepting-
tunnels driven into water-bearing strata are employed in
many parts of the world. In California, water is thus
obtained for irrigation purposes, from formations consisting
of boulders and sand. The town-supplies of Naples and
of Seville are collected by galleries deep below the surface.
The supply of Constantinople, again, is augmented by
means of the open joints of a tunnel through which the
water percolates into the aqueduct.

125. Underground galleries constructed parallel to the
banks of streams or lakes, to intercept the flow of the land-
water on its passage to such outlets, form a common type
of collecting-works. The rate of pumping from such
galleries, if greater than the infiltration of land-water into
them, induces a reverse flow from the river or lake through
the intervening strata, of an amount that depends upon the
extent by which the water-level in the gallery is lowered,
and upon the porosity of the strata. Thus we have a
second important class of tunnel-collection, in which the

* Mansergh, 'Lectures on Water Supply, &c.,' p. 65—Chatham, 1882.

works, termed in this case "filter-galleries," collect but little land-water, and serve mainly to abstract surface-water from lakes and streams, filtered to a certain extent by passing through the intervening ground.

Toulouse is thus supplied from the river Garonne. A good example of such works occurs at Lowell, Mass., where the main filter-gallery, situated alongside the river Merrimac, is 1300 feet long, 8 feet in width, and the same in height. The walls are formed of rubble masonry, and the roof by a brick arch, the level of which is approximately that of the dry-weather flow of the river. The water enters the gallery through the floor, which is covered by a sheet of coarse gravel, 1 foot thick.

The yield of springs is sometimes increased by the introduction of a puddled-clay wall, to check the passage of the land-water at that place, thus forming an artificial counterpart of the natural condition illustrated in *Fig. 1*.

126. A method of interception much practised in Holland for collecting water for domestic supply from the sand-dunes near the coast, is to intersect the dunes by canals, into which the water from the sand drains. The fine sand of which the dunes are composed, forms a natural reservoir for rain-water ; so that, even during long droughts, the water stored in it continues to drain into the canals, and the supply is maintained in an inexpensive manner. An amount equal to between 30 and 50 per cent. of the mean rainfall may be collected from these dunes, which absorb in a remarkable degree the moisture contained in the atmosphere. Instead of open channels, especially when the surface of saturation lies low, pipes with open joints are frequently laid, surrounded by shells, a modification which, owing to the conduit being closed, prevents the development of vegetable and animal life. The water thus collected in the pipes is led to a well, whence it is

raised by pumping.   Waterworks at the Hague and at
Amsterdam illustrate this principle.

On some of the plains of Flanders, where the surface
of saturation is near the ground-level, instead of canals,
numerous shallow wells are sunk.   The water is led from
these wells by siphons to a central well, whence it is
pumped to a service-reservoir.   Ejectors are used to
exhaust the air from the siphons when required, and the
water is by this means collected with an expenditure of
little power.

127. Pumping from deep wells is, owing of course to
geological conditions, the most general method of water-
supply in this country for the smaller towns and villages.
All water pumped from wells means so much abstracted
from the drainage system of the district, as an equivalent
quantity would otherwise have gravitated to the natural
outlet of the basin.   It may be, however, that the water is
returned after use to the catchment-area from which it is
taken, in a more or less polluted state, and at a lower level,

The subject of well-sinking may be conveniently divided
into two parts : (1) Geological considerations of the most
suitable sites for wells ; and (2) The methods adopted for
sinking or boring them.

128. (1) The choice of a proper place for boring is a
matter upon which depends the success of the entire
venture.   To determine it, geological features should be
considered conjointly with the physical configuration of the
district ; and the dip and strike of the strata, and the
direction and extent of faults, if any, should be closely
examined.   The most favourable position for a well is in a
large synclinal basin of high rainfall—the outlier consisting
of permeable rocks, sand, or gravel, overlying an impervious
stratum of quaquaversal dip.

Success or failure in boring frequently depends upon

the existence of fissures or faults. In chalk, which holds
the water it contains so retentively that it can only be
drawn from it at a slow rate, a bed of flints or a line of
fissures presents an extensive surface from which the water
is yielded comparatively freely. Mountain limestone
affords another example of this, and if a fissure be cut by
boring, a good supply may often be obtained from it.
Permeable strata lowered by faulting so that on the "down-
throw" side they make contact with impervious beds—if
the catchment-area is large and other conditions are
favourable—may be expected to afford, in the neighbourhood of the impervious barrier thus formed, a satisfactory
yield, for a natural underground reservoir is there formed.
Dykes of intrusive igneous rock intersecting permeable
strata produce the same effect as the faults just mentioned,
and borings made on the sides nearest the outcrop of the
strata intersected would most likely be successful ; whereas
the probability of finding much water on the other sides
of such dykes would be remote. On the other hand, a
joint filled with the rocky detritus of the surrounding
beds, might be best approached by boring on its lower
side, for it would act as an intercepting drain to the whole
of the catchment-area whose outcrop is situated on its
upper side.

The probable yield of a bore-hole or well can only be
practically estimated by comparison with similarly situated
structures ; because, in addition to the rainfall, the percolation and the absorption, the computation of which alone
forms a sufficiently difficult problem, the question becomes
so highly complex by considerations of the amount of water
leaving the strata by springs and fissures, as to baffle all
attempts to answer it directly. Where water lies in beds
situated beneath impervious strata, under a hydrostatic
pressure greater than that due to the depth of the beds

below the ground, a bore-hole carried down to tap them produces an artesian well, through which the subterranean waters rise naturally to the surface (§ 25).

The size of bore-holes is of more importance in artesian than in ordinary wells whence water is raised by pumping. In the former case, however great the available amount of water may be, the supply will only be that which the excess of the pressure at the bottom of the bore-hole over that due to the column of water in it, can force through a tube of the length, diameter and roughness of the bore-hole, according to ordinary hydraulic laws. Whereas, in the latter case, where artificial power is employed, the height to which the water naturally rises may be reduced by pumping at a greater rate ; and the percolation into the bore-hole may be correspondingly increased under the additional head thus produced. This is well exemplified in the Grenelle and Passy artesian wells, in the Paris basin, which pass through the Gault Clay into the Lower Green-sand. The former, 1795 feet deep, yielded on its completion, in 1841, about 750,000 gallons per day, through a 7-inch tube. Part of this tube having collapsed, it was replaced by one 4 inches in diameter, and the supply became reduced to about 200,000 gallons per day. The Passy well, 1939 feet deep, and 3 feet 3 inches in diameter, yielded on its completion, in 1861, nearly 4,500,000 gallons per day ; but owing to gradual obstruction of the bore by sand, this quantity fell to 1,300,000 gallons per day.[*]

A remarkably successful example of an artesian well [†] is a 4-inch bore-hole at Bourne, Lincolnshire, made in 1856 through Oolitic strata. At a depth of 92 feet below the surface, passing through hard compact rock, the boring tool suddenly fell 2 feet into a fissure, from which 500,000 gallons

[*] *Revue Scientifique*, vol. xli. 1888, p. 241.
[†] Minutes of Proceedings Inst. C E., vol. lxxv. p. 245.

of water per day rose under sufficient pressure to supply the town without the application of artificial power. Three bore-holes at Wilsthorpe,* near to Peterborough, through Oolitic strata, each 7 inches in diameter for a depth of 53 feet 6 inches, of which the last 27 feet is a bed of clay, pierce water-bearing rock for a further depth of about 12 feet, with a diameter of 5 inches. The combined yield of the wells exceeds 2,000,000 gallons per day. The bore-holes are lined with solid-drawn copper tubes, and at the top of each is fixed a valve to regulate the supply and prevent waste.

129. Wells and bore-holes in sandstone and other comparatively homogeneous rocks in which percolation is uniform, require to be of larger size than is necessary in chalk or limestone, where the supply is obtained from natural collecting-galleries in the form of beds of flints, open joints or fissures; because the percolation through a given area of the rock is determined simply by the differential head at every point. It is therefore usual to increase the collecting surface in such wells by driving tunnels at the bottom.

It almost invariably happens that the yield of ordinary wells falls off after pumping has proceeded for some time. The water from the so-called cone of depression (§ 27) becoming exhausted, the supply has to drain from greater distances, and consequently reaches the wells under smaller pressure and at a reduced rate. The clogging of the porous structure of sandstone or similar rock has a further effect in causing the supply to gradually diminish. In chalk or limestone the reverse of this may happen. For the fissures supplying wells in these formations are apt to become enlarged, owing to the solution of the calcareous rock by the dissolved carbonic acid contained in the

* Minutes of Proceedings Inst. C.E., vol. ci. p. 218.

percolating water ; their resistance to percolation is thereby reduced, and the flow through them naturally increases.

130. The most important water-bearing formations in this country are the Chalk; the Lower Greensand, lying between the Gault and Wealden Clays of the Cretaceous strata ; and the Triassic Sandstone, especially the Bunter series.   Water is also obtained in considerable quantity from the Oolitic, Permian, Millstone Grit and some other less important formations.

131. In the extensive Cretaceous districts of England are many wells bored into the Chalk, for general use as well as for private supplies—the latter especially in the London basin, in which the level of the water has been artificially lowered until it is now considerably below the mean level of the river.   One cubic foot of chalk can absorb 2 gallons of water, whilst one cubic foot of sandstone holds but little over ¾ gallon.   Chalk is, however, highly retentive, and in order to obtain a good supply from it, some of the numerous beds of flints occurring in the upper Chalk must be intersected, or cavities and lines of fissures, which form natural reservoirs into which the water slowly percolates.

From the results of experiments and investigations in 1851, Prof. D. T. Ansted arrived at the conclusion that chalk is so absorbent and retentive of moisture that the fissures in it tend rather to supply water to the rock than to collect any therefrom.   Evaporation proceeds rapidly from its surface, water being transmitted upwards by capillary action to supply the loss.   Otherwise, water does not pass out from solid chalk by percolation; and the available water in this formation consists only of so much as exists in crevices and fissures, by which it is fortunately intersected in a remarkable degree.*

* Ansted, ' Water and Water Supply,' p. 84—London, 1878.

The West Kent Water Company derives a large supply from the Chalk in the neighbourhood of the fault that traverses the Deptford district. * It has been often observed that little water is generally obtained from chalk which is buried by a considerable thickness of other strata. This has sometimes been attributed to the closing of the fissures by the superincumbent pressure. A more probable explanation is that the carbonic acid contained in solution in rain-water, augmented by that derived from decomposing vegetable matter during its passage through the soil, becomes combined in the upper strata before the water reaches the chalk. There is therefore no agent at work forming holes and enlarging fissures in the latter. Certainly, if at any previous time the surface of chalk had been exposed when newer beds were deposited upon it, there would be a tendency to fill up fissures already formed in it. It is therefore generally stated that the outcrop of the Chalk is the most likely place to find water by boring.

The Oolitic rocks are almost as absorbent as the Chalk, whilst percolation takes place through them much more freely. They would rank among the finest water-bearing formations, were it not that the very porosity of their structure renders them peculiarly liable to pollution by infiltration of surface impurities. †

132. The lower Trias is but little fissured, and the supply from wells in it depends on the percolation through the rock and along the joints in it. The water afforded by it is usually of excellent quality, if somewhat hard for ordinary consumption ; though where the Triassic marls overlie the Bunter sandstone, the water obtained from the formation

---

* Colburn and Maw, ' The Waterworks of London,' p. 67—London, 1867.
† De Rance, ' The Water Supply of England and Wales,' p. 41—London, 1882.

frequently contains large quantities of gypsum and rock salt in solution, and is then unsuitable for domestic supply —a cause that has led to the rejection of such water after the expense of boring or sinking wells in this formation has been incurred.

133. Wells of large diameter sunk through soft earth are generally lined with brickwork set in cement, carried upon a timber curb which sinks as the excavation proceeds. If lined with cast-iron cylinders, the bottom one is provided with a cutting-edge of slightly larger diameter than that of the cylinder, and the ground may be removed from the interior by a grab, or by the ordinary pick-and-shovel process. It is desirable to fix a timber framing, in order to guide the cylinders vertically for the first 15 or 20 feet. The frictional resistance opposed by the earth to sinking may at great depths amount to $\frac{1}{4}$ ton per square foot of surface. When the cylinders stick, a small charge of dynamite exploded at the bottom will often cause them to re-commence their descent.

134. Wells of considerable depth in rock can now be bored more cheaply than they can be sunk in the old way, even when so large as 15 feet in diameter. A boring of such size as this requires, however, a pilot bore-hole about 6 feet in diameter to be driven first. Two wells of 6 feet diameter and 100 feet deep, bored near to Otterbourne for the Southampton Waterworks,[*] form interesting examples of recent well-boring by the percussive method in this country. The boring-tool employed consisted of a flat wrought-iron cross-head, to which three wrought-iron chisels with steel points 2 feet broad were attached by cottars, forming a trident. The two outer chisels were furnished with short iron guides formed so as to act as rimers, and were further attached to the cross-head by keep-chains in case of

[*] Minutes of Proceedings Inst. C.E., vol. xc. p. 33.

accident. The weight of the tool was 1¼ ton. The boring-rods were of iron, 3 inches square, coupled in 10-foot lengths with screwed joints, and were suspended to a 2-inch lifting-cable with a swivel and shackle. A steam-winch was kept running constantly, round the drum of which a few turns of the lifting-cable were coiled ; when the loose end of the cable was hauled taut the rods were raised, and on its being released they dropped instantly. A rotary motion was imparted to the rods and boring-tool by manual power.

The miser used was made of ⅜-inch boiler-plate stiffened by angle-bars and plates. Each half of the lower portion was made tapering with a helical twist, the upper edge of one plate being brought nearly vertically over the lower edge of the other plate, and between the two a hinged flap was hung, to retain the debris when forced into the miser. The edge of each plate was furnished with steel dog-tooth cutters and a set of points ; cutters were also bolted to it to finally rime out the wells. This tool weighed 1¼ ton, and was rotated by manual power. The second of the two wells was sunk at an average rate of more than 5 feet per day.

These wells were bored by Messrs. Le Grand and Sutcliffe, who have favoured us with further information as to a recent bore-hole executed by the same method at New Lodge, Windsor. This boring was commenced with a diameter of 7¼ inches in the expectation that water would be obtained from the chalk at a depth of 250 to 300 feet. Failing this, the rock was pierced to a depth of 1120 feet, the bore-hole being lined with 3-inch tubes to that point and finished with a still smaller diameter, to a total depth of 1242 feet, in the Lower Greensand. Had it been known at the outset that the boring would have to be carried to so great a depth, larger diameters would have been employed.

135. In sinking the shafts of mines, it is a practice on the Continent to first bore a well about 6 feet in diameter, which is subsequently widened out to the full diameter by a tool consisting of an inverted T-shaped forging carrying a set of vertical chisels, which diminish in length towards the outside so that the detritus falls into the smaller central bore-hole. The "shell-pump" employed to lift the debris in this work consists of an iron cylinder divided into four compartments by transverse partitions, each of which has its own valve, and of less diameter than the central pilot-hole in which it works.

Borings of smaller diameter in rock are effected much in the same way with a single chisel. If in sand, a shell-pump is used ; and a special auger is employed for piercing clay.

136. An important modification of the above method of boring is that afforded by the diamond drill, which extracts a solid core, and is consequently of great utility when it is desired to examine the character of the strata bored through. It may be employed for cutting bore-holes up to 2 feet in diameter. In the larger sizes, the working crown in which the diamonds are set is of wrought-steel, about 2 inches thick. This crown is screwed into a core-tube which is attached to hollow rods, and the boring is effected by the rotary motion of the crown imparted to it through the rods by machinery at the surface. The diamonds are of the dark amorphous variety (the *carbonado* of Brazil), about the size of small beans. They are brazed to steel plugs, which fit into holes drilled in the crown, about $1\frac{1}{2}$ inch apart.

The core-tube, to which the crown is screwed, is of wrought-iron, 30 feet long ; the rods, made of drawn-steel about $3\frac{1}{2}$ inches in external diameter and $\frac{3}{8}$-inch thick, coupled together in 5-foot lengths, are connected with a plate at the top of the core-tube. The latter, extending upwards 5 feet, is open at the top so as to receive the fragments of rock washed up by a stream of water which is forced down through the hollow rods, or falling from the sides of the bore-hole. The cross-head from which the rods are driven slides between the rotating vertical shaft and a fixed cast-iron pillar, and the rotary motion is given to the rods by means of a spur-wheel, for which the cross-head forms a supporting pedestal. This gears into a spur-wheel sliding over a broad feather-key on the vertical shaft below the level of the cross-head, in order to allow for the downward motion of the rods. A weight is arranged to counterbalance the rods as the boring becomes deeper, so that the pressure on the crown does not ordinarily exceed half a ton, depending upon the diameter of the tool. The crown rotates at 40 to 200 revolutions per minute, according to the rock in which it works. After a 30-foot length has been drilled, the core-tube and crown are lifted to the surface, the crown is unscrewed, and an extractor is attached to the core-tube. The rods are raised by shear-legs and tackle, and are detached in 40-foot lengths. A 20-I.H.P. steam-engine suffices for the work under ordinary circumstances.

Plant such as that described was used by Messrs. Docwra & Son, in sinking two borings at Northampton ; * the first, a bore-hole 850 feet deep, was sunk in the expectation of reaching water-bearing strata of the Triassic formation. This formation, however, proved to be absent at that place, mountain limestone being encountered. The rate of boring was between 6 and 12 feet in twelve hours, though a much higher rate may be attained for short periods. Simpler forms of the machine are used in smaller borings.

* Minutes of Proceedings Inst. C.E., vol. lxxiv. p. 270.

137. The method of suspending the boring apparatus by a rope, which obviates the difficulties inseparable from the use of coupled boring-rods, was, after its early introduction, abandoned owing to the difficulty of boring vertically; but more recently Messrs. Mather & Platt, of Manchester, have succeeded in rendering the method practicable. They use a flat hempen rope, $\frac{3}{4}$ inch thick and $4\frac{1}{2}$ inches broad, which is wound round a large drum driven by a steam-engine; in their later practice a round rope has been used. On leaving the drum, the rope is guided by a pulley so as to lap over another pulley carried in jaws at the end of a piston-rod, the piston of which reciprocates in a vertical steam-cylinder. When the boring-tool suspended by this rope has been let down the hole, the rope is clamped in position near to the drum, the piston being at the bottom of the cylinder. Steam- and exhaust-valves, for effecting the reciprocation of the piston, are worked by tappets moved by the piston-rod. The steam-valve is placed at the bottom of the cylinder and the exhaust-valve 6 inches above it, so that there may be a cushion of steam under the piston when it descends. The valves being actuated thus, an arrangement is required to set the piston in motion. This is effected by a small pipe constantly delivering steam through the bottom of the cylinder, by means of which the piston is slowly raised sufficiently to cause the main steam-valve to open—when the piston is driven to the top of the cylinder, raising the boring-tool twice the height of its stroke. The exhaust-valve is then opened, and the piston descends by its own weight, allowing the boring-tool to fall upon the bottom of the hole. This tool consists of a wrought-iron bar, to the bottom of which is attached a cylindrical block, in which steel cutters are inserted and secured by nuts, so that they may be easily removed to be sharpened or renewed.

Above the block one or more cylindrical iron castings are provided to act as guides.

A definite rotation is given to the tool by two cast-iron collars, cottared to the top of the bar. On the lower edge of the top collar, and on the upper edge of the bottom one, are formed deep ratchet teeth of the same pitch. Between these two collars a brush with corresponding ratchet teeth on both sides slides on the bar, and is attached to the rope, the teeth on its upper side being one-half the pitch in advance of those on its lower side. As the tool descends the upper teeth of the bush are in contact with the teeth of the upper collar, and when the tool reaches the bottom of the bore-hole, the bush falls on to the lower collar, causing the rope to twist through half the pitch ; when the rope ascends, the upper teeth of the bush again engage with those of the upper collar, which twists the rope through another half pitch. The rope then turns the tool through the same angle in untwisting at each complete stroke.

With this boring-machine about 24 blows per minute can be given, and it can be worked by two men, one of them attending to the engine and changing the tools, and the other emptying the " shell-pump," or apparatus by which the comminuted rock is collected and lifted from the bottom of the bore-hole. This consists of a cylindrical iron barrel, a little smaller in diameter than the bore-hole, containing an indiarubber valve at the bottom, and sus-pended by a long link from the same rope as that employed to carry the boring-tool, after the latter has been drawn up to the surface and disconnected. The debris enters the shell by the flap-valve at the foot, which is loose to facilitate emptying. In use, the pump is lowered down the bore-hole by the rope, and is then raised and lowered two or three times through the range of the working-stroke of the boring-tool, after which it is drawn

K

up and emptied; an operation that may be conveniently repeated every twenty minutes in this method of boring. In traversing beds of clay, a cylindrical grab is sometimes used for the larger sized bore-holes.

A bore-hole at Bootle, Lancashire, 1302 feet deep and varying in diameter between 26 inches in the upper and 20 inches in the lower part, made with this apparatus in 1877–78, occupied 17 months in construction. The deepest bore-hole yet made by Messrs. Mather & Platt is at Lackenby, near Middlesbrough, 1806 feet deep; the diameter varying between 12 and 8 inches. In this and in other borings a rate of 40 feet per day has been attained in sandstone. In a boring made through chalk at Bushey, Herts, in 1885, with one of these machines, the diameter being 18 inches in the upper portion and 15 inches for the lower 180 feet, the unlined part, 617 feet in depth, was bored in 38 working days, including stoppages for repairs —an average of 16 feet per day. The greatest depth bored in one day was 30 feet.*

138. In boring through soft strata, to keep the hole open; or to prevent the ingress of surface-water, or loss in the case of artesian wells, the bore-hole must be made larger than is otherwise necessary; in order that wrought- or cast-iron tubing, generally screwed together with flush joints, may be introduced into it. The bottom tube has a sharp cutting-edge of hard steel. The boring and the introduction of the tubing proceed simultaneously, except during the short time occupied in inserting a new tube. When the tubes become earth-bound, a cap is screwed on the top one, and they are either driven with a mall, or forced down by screw-jacks, in small, or by hydraulic jacks in larger bore-holes.

To prevent surface-water, which may find its way down

---

* Minutes of Proceedings Inst. C.E., vol. xc. p. 21.

the outside of the tubes, from entering the bore-hole at the bottom of the portion thus lined, a hollow conical plug is sometimes forced down to that place. This is designed to form a water-tight joint between the tubes and the lower, impervious stratum. In small bore-holes, however, when it has to be fixed from the top, its efficacy is doubtful. The best means of preventing this action, in wells of such size as to allow of men working inside them, is to inject cement grout under pressure round the cylinders through holes which are afterwards tapped and plugged ; this possesses the further advantage of preserving the cylinders from decay.

139. For small supplies, the tube-well forms a useful apparatus. It consists of strong wrought-iron pipes from 1 to 2 inches in diameter, screwed together. The bottom pipe terminates in a point, and is perforated with small holes. The point is driven into the ground by a falling weight, as in pile-driving, and additional lengths of pipe are screwed on as it descends. When this kind of well is sunk through fine sand, the work may be performed expeditiously by forcing water through the tubes at a velocity of 10 to 15 feet per second.

140. By the methods described, wells have been bored to depths of several thousands of feet. The deepest bore-hole in the world was made at Schladebach in Prussia, and is 5734 feet deep.* The diameter is 11·2 inches at the top, and decreases to 1·25 inch at the bottom. It was bored with the diamond drill. The temperature increased at the rate of 1° C. for every additional 120 feet in depth. A boring 5 inches in diameter and 4500 feet deep, completed in 1891, at Wheeling, West Virginia, showed a rise of temperature of 1° C. for every 130 feet increase of

* 'Zeitschrift für das Berg-, Hütten- und Salinen-wesen,' vol. xxxvii. Abhandlungen, p. 172—Berlin, 1889.

depth,* the rate of increase being rather more rapid at the greater depths. These rates of increase of temperature with the depth are somewhat less than those observed in deep bore-holes in England—the average increment in our latitude having been observed to be about 1° C. for every 100 feet in depth.† This question assumes some importance in respect of water supplies sought by this means in the lower formations of the earth's crust ; where temperatures obtain at which it is generally undesirable to supply water for domestic use, or to store it for long periods.

141. Hitherto we have considered only the means by which access is obtained to natural sources of water supply, which, in certain cases, can be rendered available for use without the resort to any power for purposes of conveyance and distribution other than that afforded by the action of gravity. Artificial power and appliances have, however, so frequently to be employed to raise water from lakes, streams and wells, to elevations from which the supply may flow and be distributed by the uniform and continuous gravitating impulse of the water, that we proceed to discuss them in some detail. Motive power for pumping purposes is supplied by steam, by water in motion or under pressure, and by the wind ; the importance of the first being, however, much greater than the latter two, because it may be rendered available wherever and whenever required. The pumps employed are generally of the reciprocating type for high lifts, and of the rotary type for low lifts.

142. The three varieties of the former type of pump in most common use are : 1, the plunger-pump ; 2, the bucket-pump ; 3, the bucket-and-plunger pump.

---

* Report of the British Association, 1892, p. 129.
† Ramsay, 'Physical Geology and Geography of Great Britain,' p. 50—London, 1878.

1. The plunger works through a stuffing-box, and acts entirely by displacement. The wear is consequently slight. It may be either single- or double-acting.

2. The bucket-pump is subject to more wear, as the bucket must make a water-tight joint with the barrel in which it works, containing a valve or valves through which all the water passes. This pump is necessarily, in its simple form, single-acting. It possesses the advantage of having no gland, and may be worked whilst completely submerged.

3. In order to render the bucket-pump double-acting the pump-rod is thickened out to form a plunger, occupying about half the volume of the working barrel, and passing through a stuffing-box. This forms the class of pump in most common use. The delivery is continuous on the down stroke, as well as on the up-stroke, and, with fewer valves than in the double-acting plunger-pump, it may be worked at somewhat higher speed than the latter.

143. Pump-valves are formed by either a number of small orifices, furnished with indiarubber or leather flaps, or with gun-metal disks ; or they may be of the double- or multiple-beat, or the annular-grid types. An example of the first mentioned type of valve is formed of indiarubber or leather disks backed by gun-metal, sliding on fixed vertical spindles, with stops to prevent them from rising too high. The gratings on which the valves work are machine-faced. In the other variety, the valves are disks of gun-metal sliding on spindles, with indiarubber stops or spiral springs behind them to assist their closing-action.

The double-beat valve is so well known as hardly to need description here. Its seat has gun-metal, gutta-percha or hard wood faces.

The annular-grid valve has a seat consisting of a horizontal gun-metal grating with concentric series of openings, over which works the valve, formed by a flat plate, in which similar series of concentric apertures are arranged so that the solid part of the plate covers the openings of the grating. The valve slides on a central vertical spindle. In a valve of somewhat similar construction, termed the "bee-hive," the perforated metal rings forming the valve-seat are separate, and arranged in tiers of diminishing diameter, telescope fashion, the valves being formed by loose indiarubber disks laid over the several grids, and kept in place by the vertical portion of the adjacent smaller rings. This type has proved effective and durable in practice.

144. Valves should be designed with seats narrow relatively to their apertures, in order that the surface of the moving parts exposed to the water when the valves are lifted may not be much greater when they are on their seats. This prevents the valves from being shot up violently as soon as they are opened, and so avoids the shocks that would otherwise occur from that action. The lift of the valve should be small, and the total opening should be large, opposing little resistance to the passage of the water through it. The valve will then, after rising, resume its seat quickly, and prevent reversal of the current of water, and the consequent slip. To avoid the effect of slip, the valves in plunger-pumps have sometimes been actuated mechanically, but hitherto this modification has not received extensive application.

145. Uniformity in the delivery of pumps is always desirable, and is essential when water is pumped directly into distribution pipes, without the intervention of a reservoir or stand-pipe. It depends primarily upon the manner in which the pumps are driven. In a double-acting pump it may be ensured by causing the plunger to move at a uniform speed : but if the motion is governed

by a fly-wheel, the delivery is nearly harmonic. The former condition is effected in pumps of the "Worthington" type. In the latter case, the uniformity of delivery depends upon the arrangement of the pumps and the phase of their motion relatively to each other.

To reduce pulsation in delivery and variation of the pressure in the main, an air-vessel, consisting of a strong iron chamber communicating with the delivery-main, forms an important accessory to every pumping installation. It possesses the further advantage of preventing shocks arising from change of momentum of the water put into motion or stopped by the opening and closing of the valves. To be most effective for this purpose, an air-vessel should be situated near to the pump on the delivery-main, and connected with the latter by a branch pipe of large diameter. If a large mass of water is put into motion at each stroke in the suction-pipe, the latter also should be provided with a foot-valve, and with an air-vessel to effect a gradual arrest of the water and prevent undue impact on the pump-valves. A chamber for this purpose is sometimes formed round the pump-barrel or the valve-chamber.

146. Assuming the motion of the pumps to be harmonic, the quantity of water delivered through each barrel to any point θ of the stroke of the piston is $\dfrac{A\,L}{2}\,(1 - \cos\overline{\theta - a})$; where L is the length of stroke and A is the area of the piston; $a = 0$ for the first pump-barrel, and 90°, 120° or 180° for the other barrels, according to the angles which the respective cranks make with the first one; and θ is an angle varying from 0° to 180°.

The rate of pumping in each barrel at any instant is therefore

$$\frac{A\,L}{2}\,\sin\,(\theta - a);$$

and the average rate throughout the stroke is for

a single-acting pump ($a = 0$).. .. .. $\left\{ \dfrac{A\,L}{2\,\pi} \right.$

2 single-acting pumps ($a = 90°$ or $180°$), or for a $\left\{ \dfrac{A\,L}{\pi} \right.$
   double-acting pump ($a = 180°$) .. .. ..

3 single-acting pumps ($a = 120°$) .. .. .. $\left\{ \dfrac{3\,A\,L}{2\,\pi} \right.$

4 single-acting pumps, or for 2 double-acting $\left\{ \dfrac{2\,A\,L}{\pi} \right.$
   pumps ($a = 90°$ and $180°$) .. ..

The maximum excess of the delivery above the mean is obtained by finding the points of intersection of

$$y = \frac{A\,L}{2\,\pi}\,; \; y = \frac{A\,L}{\pi}\,; \; y = \frac{3\,A\,L}{2\,\pi}\,; \; y = \frac{2\,A\,L}{\pi} = \eta, \text{ say,}$$

with $\qquad\qquad y = \Sigma\,\dfrac{A\,L}{2}\,sin\,(\theta - a),$

—where $a = 0$ for the first pump-barrel ; $a = 90°$, $120°$, or $180°$ for the others respectively ; and $\Sigma\,\dfrac{A\,L}{2}\,sin\,(\theta - a)$ denotes the curve resulting from the combination of the harmonic curves representing the rates of the several pumps—and integrating the expression

$$y = \Sigma\,\frac{A\,L}{2}\,sin\,(\theta - a) - \eta$$

between the limits of the first and other intersections.*

The ratio that this excess bears to the total delivery per stroke is, in

a single-acting pump ($a = 0$) .. .. .. $55\cdot0$ per cent.
2 single-acting pumps ($a = 90°$).. .. .. $35\cdot0$ „
2 single-acting pumps, or a double-acting $\left.\right\}10\cdot5$ „
   pump ($a = 180°$) .. ..

---

* A graphical solution of this question has been given by the late Mr. W. E. Rich, *vide* Minutes of Proceedings Inst. C.E., vol. lxxviii. p. 24.

4 single-acting pumps, or 2 double-acting pumps ($a = 90°$ and $180°$) .. .. $\Big\}$ 1·1 per cent.

3 single or double-acting pumps ($a = 120°$)   0·3 „

147. A simpler, if somewhat less obvious, method of arriving at these ratios is afforded by the following :

The delivery D up to any point θ of the stroke of the pistons is $\dfrac{A L}{2} \Sigma (1 - \cos \overline{\theta - a})$; and the mean delivery M up to any point θ of the stroke of the pistons is $\dfrac{A L}{2} \cdot \dfrac{m \theta}{\pi}$; X is $D - M$; and X_0 is the value of X when $\theta = 0$;

where $a = 0°$, $90°$, or $180°$, according to the phase of the particular piston considered,

and $m = 1, 2, 3$ or 4, according as the pump is of single-acting, double-acting, three-throw or four-throw type.

The excess of D above M is given by

$$X = \frac{A L}{2} \left\{ \Sigma (1 - \cos \overline{\theta - a}) - \frac{m \theta}{\pi} \right\} - X_0,$$

and this has a maximum value when

$$\Sigma \sin \overline{\theta - a} = \frac{m}{\pi}.$$

E.g. For a three-throw pump,

$$D = \frac{A L}{2} \left\{ \frac{1}{2} + (1 - \cos \theta) \right\},$$

$$M = \frac{A L}{2} \cdot \frac{3 \theta}{\pi},$$

and

$$X = \frac{A L}{2} \left\{ \frac{1}{2} + 1 - \cos \theta - \frac{3 \theta}{\pi} + 0.009 \right\}.$$

This last expression has a maximum value when

$$sin\,\theta = \frac{3}{\pi};$$

i. e. when

$$\theta = sin^{-1}\frac{3}{\pi}.$$

Therefore the required ratio

$$\frac{Max.\ value\ of\ X}{Total\ delivery\ per\ stroke} = 0\cdot003.$$

148. The size of air-vessels for ordinary working is arrived at by multiplying the fractions that represent these percentages by the delivery per revolution, and dividing by the fraction representing the limiting value of the variation of pressure in the main.

For example, for a variation not exceeding $\frac{1}{50}$th of the average pressure, the above ratios multiplied by the delivery per revolution and by 50 give the capacity required in the several cases. When two or more pumps are employed, it is desirable to make the air-vessel large enough to enable pumping to be continued without excessive fluctuation of pressure in the event of one of the pumps breaking down.

The ratio of the maximum excess of the delivery over the mean delivery with one pump out of work is, in the case of

2 single-acting pumps ($a = 180°$) .. .. .. 55 per cent.
3 single-acting pumps ($a = 120°$) .. .. .. 27 „
4 single-acting pumps ($a = 90°$ and $180°$) .. 17 „

The variation of pressure permitted in the main may under such exceptional circumstances be greater than in the former case, say twice as much. Therefore to find the capacity of the air-vessel required, the reduced delivery would be multiplied by these fractions and by 25.

149. The supply of air, gradually dissolved in the water, is replenished in the air-vessel, either by a small force-pump or by means of a " snifting valve." This latter ingenious device consists of a chamber provided at the top with an inlet-valve communicating with the atmosphere, and a pipe with a reflux-valve leading into the air-vessel ; also furnished at the bottom with a pipe communicating with the space between the suction- and delivery-valves of the pump. During the suction-stroke, air is drawn into the chamber, and during the delivery-stroke it is filled with water from the pump-barrel, which expels the air, forcing it into the air-vessel. The inlet air-valve and the pipe from the chamber to the pump-valves are of such relative size that in the suction-stroke the pipe is not emptied of water ; air is thus prevented from entering the pump.

150. Instead of an air-vessel, an iron or steel stand-pipe is often used, in which the water-level oscillates about a mean position corresponding with the average pumping-head employed. Its height above the most elevated point in the district to be supplied is equal to the head necessary to force the supply through the distributing- and service-pipes. Stand-pipes are generally situated near the pumps, and are designed of such dimensions that the excess of delivery in each stroke above the mean may not raise the height of the water in them more than corresponds with the allowable variation of pressure in the mains. In the United States it is a common practice to employ stand-pipes of huge dimensions—we do not allude to those which may be useful as service-tanks (§ 266) : that at Bloomington, Illinois, is 8 feet in diameter and 200 feet high.

151. It will be observed that, as the flow of the water through reciprocating pumps is necessarily discontinuous, the speed at which they can be worked depends upon the length of the pump-rods, the valve-areas, and the provision

made, by air-cushions and otherwise, for absorbing that portion of the momentum of the water which is due to its varying motion in the suction-pipes and through the pumps.

The fundamental principle of the design of these pumps is that the rate at which effective work, measured by the water raised, is performed, may approximate closely to the power applied to the machines—the ratio of the former to the latter quantity is termed the "pump efficiency." The causes which tend to diminish the efficiency of reciprocating pumps are : the inertia of the water, and the friction of the suction- and the delivery-pipes, and of the pump-barrels ; resistance at the valves ; the periodic stoppage of the motion in the suction-pipe, and slip ; variation of the motion in the pump-barrels and delivery-pipe ; and friction of the working parts.   The first mentioned group of these causes are precisely analogous to the phenomena presented for consideration by the motion of water in a gravitation main (§ 318) ; the remainder have been discussed, as far as concerns the steps necessary to reduce their effects in practice, in §§ 143–150.

In favourable circumstances, with pumps of any considerable size, the effective work approaches within 5 per cent. to the power applied.

If E is the efficiency of a pump,

P „ power applied to it in foot-pounds per minute,

V „ volume of water delivered per stroke,

n „ number of strokes per minute,

H „ lift in feet from the suction to the point of delivery,

h „ head in feet absorbed by the causes mentioned above,

and w „ weight in pounds of a cubic foot of water,

$$P = n V . w (H + h),$$

and $$E = \frac{n V . w H}{P}$$

$$= \frac{H}{H + h} .$$

152. When large quantities of water are to be raised through a moderate lift, the centrifugal pump is conveniently employed, and, the flow of the water through it being continuous, may be driven at a high speed. It may be described as essentially a reaction turbine, which, instead of being worked by water under pressure, is rotated by mechanical power, and lifts a column of water.

To obtain the best effect, a centrifugal pump must work at the definite rate for which it has been designed. Any increase or diminution of that rate acts prejudicially upon its efficiency. The water is drawn through a suction-pipe into the centre of a rotating wheel, and, in the simplest form of the machine, is thence flung by centrifugal action into a circular chamber from which it flows directly to the delivery-pipe. In the improved type of the pump, *Fig. 20*, an inner chamber, within the pump-case, surrounds the wheel; and the water, set in motion by the curved vanes of the latter, rotates within the chamber in a free vortex. Part of the energy of this whirling motion of the water is utilised in augmenting the pressure in the pump, and its efficiency is correspondingly increased. A foot-

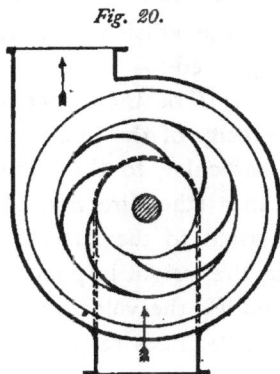
Fig. 20.

valve is sometimes used, so that the pump may be filled with water at starting—a plug being provided in the upper part of the casing, or in the delivery-pipe, for this purpose. Starting may be also effected by exhausting the air out of the casing and the suction-pipe.

153. Let Q be the quantity of water in cubic feet per second that the pump is designed to lift; H the height of the lift in feet; r_0, b_0, are the radius and breadth respectively of the opening, in feet, at the outlet-circle (*Fig. 21*); u_0 the radial velocity of flow in feet per second at radius r_0; r_1, u_1, similar quantities at the inlet circle (*Fig. 21*);

then $$Q = 2 \pi r_0 b_0 u_0,$$

neglecting the space occupied by the thickness of the vanes. In practice

$$u_0 = 0 \cdot 25 \sqrt{2 g . H}, \text{ nearly.}$$

To find the "hydraulic efficiency" (E_H) of the pump, i.e. the efficiency if all losses arising from friction are neglected :—

If T be the torque acting on the pump; ω, the angular velocity of the pump-vanes; v_0, the velocity of whirl perpendicular to the radius at the outlet-surface from the vanes (the direction of the flow at the inlet-circle being normal to the latter, there is no velocity of whirl there); w, the weight in pounds of a cubic foot of water; and V_1 and V_0 the velocities of the pump-vanes at the inlet- and outlet-circles respectively; the work done on the pump is

$$T \omega.$$

Since the velocity of whirl at the inlet from the centre is nil,

$$T = \frac{Q w}{g} v_0 r_0$$

i. e. the torque equals the moment of momentum communicated to the water passing through the pump per second.

The work done by the pump is $Q\,w\,H$.

Therefore
$$E_H = \frac{Q\,w\,H}{\dfrac{Q\,w}{g}\,v_0\,r_0\,\omega}$$

$$= \frac{g\,H}{v_0\,r_0\,\omega}\,;$$

or, since $r_0\,\omega = V_0$,

$$E_H = \frac{g\,H}{v_0\,V_0}.$$

If ϕ (*Fig. 21*) denotes the
angle that the tangent to a vane
at its outer end makes with the
tangent to the outlet-circle at
the same point, the former tangent being in the direction of
water leaving the vane relatively
to its motion at that point ;

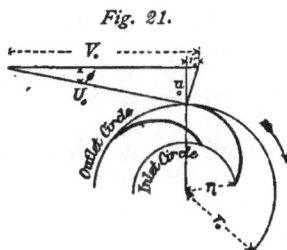

Fig. 21.

p_1, p_0, the pressures in pounds per square foot between
a pair of vanes at their inlet- and outlet-surfaces
respectively ; and

U_1, U_0, the velocity of the water relatively to that of
the vanes, at those points ;

then
$$v_0 = V_0 - u_0 \cot \phi, \qquad \text{(i.)}$$

therefore
$$E_H = \frac{g\,H}{V_0\,(V_0 - u_0 \cot \phi)}.$$

Also
$$\frac{p_0 - p_1}{w} = \frac{U_1^2 - U_0^2}{2\,g} + \frac{V_0^2 - V_1^2}{2\,g}.$$

The derivation of the first term on the right of this equation is an application of Bernoulli's theorem (§ 65), and the second term expresses the effect of the forced rotation of the water.*

Now $$U_1^2 = u_1^2 + V_1^2$$

because the flow is radial at the inlet circle,

and $$U_0 = u_0 \, cosec \, \phi,$$

therefore $$\frac{p_0 - p_1}{w} = \frac{u_1^2 - u_0^2 \, cosec^2 \phi + V_0^2}{2g}.$$

In this expression, $\frac{p_0 - p_1}{w}$, the gain of pressure in the pump, is equal to $H + \frac{u_1^2}{2g}$, the lift together with the head required to give the initial motion to the water ;

therefore $$H = \frac{V_0^2 - u_0^2 \, cosec^2 \phi}{2g}, \qquad \text{(ii.)}$$

or $$V_0 = \sqrt{(2gH + u_0^2 \, cosec^2 \phi)},$$

---

* Since the centrifugal force of each small radial cylinder of water of length dr is balanced by the difference of pressure dp on its ends, and the angular velocity ω is constant,

$$\frac{w \, dr}{g} \cdot \omega^2 r = dp \; ;$$

but $$\omega r = V,$$

therefore $$\frac{V \, dV}{g} = \frac{dp}{w}.$$

Integrating, $$\frac{V_0^2 - V_1^2}{2g} = \frac{p_0 - p_1}{w}.$$

and
$$E_H = \frac{V_0^2 - u_0^2 \, cosec^2 \, \phi}{2 \, V_0 \, (V_0 - u_0 \, cot \, \phi)} \, .$$

If ϕ is 90°, and the vanes are consequently straight,

$$E_H = \frac{V_0^2 - u_0^2}{2 \, V_0^2} \, .$$

Taking u_0 equal to $0 \cdot 25 \, \sqrt{2 \, g \, H}$,

$$E_H = 0 \cdot 47 \; ;$$

but as ϕ is diminished, the value of E_H correspondingly increases.

154. By the addition of a chamber partially enclosing the wheel, additional advantage is derived from the whirling motion of the water in a free vortex after it has passed through the vanes of the pump, an increased difference of pressure between the suction and delivery of the pump being obtained, the amount of which depends principally upon the outer radius R of this chamber. In a free vortex the total energy along each stream-line is constant, and the velocity varies inversely as the radius ;* therefore the increased difference of pressure is

$$\frac{v_0^2 + u_0^2}{2 \, g} \left(1 - \frac{r_0^2}{R^2} \right),$$

---

* In a free vortex, the total energy along each circular stream-line is the same. Therefore, differentiating Bernoulli's equation, we obtain

$$\frac{v \, dv}{g} + \frac{dp}{w} = 0 \; ; \qquad \text{(i.)}$$

and, since the centrifugal force of each small radial cylinder of water of length dr is equilibrated by the difference of pressure dp on its ends,

$$\frac{w}{g} \, d r \, \frac{v^2}{r} = dp \; ;$$

substituting in (i.)
$$\frac{dv}{v} + \frac{dr}{r} = 0.$$

Integrating
$$log \, v \, . \, r = constant,$$

$$\text{or} \quad v \propto \frac{1}{r}.$$

L

which, substituting the value of v_0 (§ 153, (i.)),

$$= \frac{V_0{}^2 - 2\,V_0\,u_0\,cot\,\phi + u_0{}^2\,cosec^2\,\phi}{2\,g}\left(I - \frac{r_0{}^2}{R^2}\right).$$

Adding this to the expression previously obtained for H (§ 153, (ii.)), the new value of E_H becomes

$$E_H = \frac{\left(2 - \frac{r_0{}^2}{R^2}\right)V_0{}^2 - 2\left(I - \frac{r_0{}^2}{R^2}\right)V_0\,u_0\,cot\,\phi - \frac{r_0{}^2}{R^2}u_0{}^2\,cosec^2\,\phi}{2\,V_0\,(V_0 - u_0\,cot\,\phi)}.$$

If ϕ is 90°, the vanes being straight,

$$E_H = \frac{\left(2 - \frac{r_0{}^2}{R^2}\right)V_0{}^2 - \frac{r_0{}^2}{R^2} \cdot u_0{}^2}{2\,V_0{}^2} ;$$

and if $R = 2\,r_0$, $E = 0.87$.   As in the former case, when ϕ is less than 90°, as it generally is in practice, the value of E is duly increased.

This investigation of the "hydraulic efficiency" of centrifugal pumps places the consideration of their efficiency on the same basis as that of reciprocating pumps. The losses owing to friction have still to be deducted from E_H above.   Since the motion of the water in centrifugal pumps is continuous, losses corresponding with those due to variation of the motion in reciprocating pumps are absent.

155. The Archimedean-screw pump, *Fig. 22*, consists of a sheet-iron cylinder with a concentric central core.   Several spirally-wound iron blades traverse the annular space thus formed.   The ends of the core are carried by journals, not shown in the diagram, revolving in pedestals, the lower end being immersed in the water to be raised ; the inclination of the cylinder, and its length, being regulated by the height

of the lift required. The pump is generally driven by spur-
or bevel-gearing attached to the outer cylinder at its upper
end, the motive power being furnished by steam, as at the
Antwerp Waterworks, by water, or by wind. It is inapplicable when the water-level on the delivery side rises
above the cylinder.

The following rules for the design of these pumps are
due to Mr. Wilfred Airy, who has paid special attention to
this machine. The quicker the spiral, the smaller should
be the inclination of the machine, in order to produce the

Fig. 22.

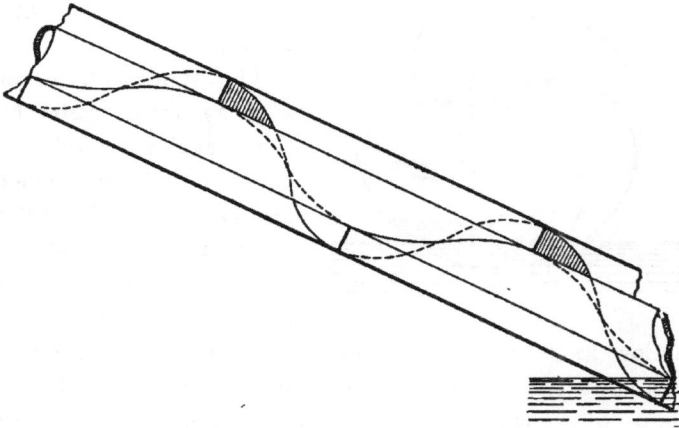

best effect. A quicker spiral will lift more water than a
slower one at an equal rate of revolution. The blades
may be conveniently formed of continuous sheets, and in
that case the machine works to the best advantage when
placed so that the acute angle made by them with the core
is downwards ; and this angle varies with the spiral.

A machine of this description, 3 feet in diameter and
40 feet long, can deliver 50,000 gallons of water per hour,
with a lift of 20 feet.

In Holland, this pump is made without an external case, consisting merely of a core or axis carrying a spiral blade that revolves close to a fixed semi-cylinder of masonry or wood. In this form, the leakage between the spiral and the envelope is a serious drawback, and the pump is inferior to the kind described above.

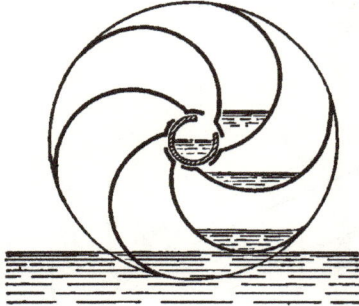

156. There are many other forms of lift-pump used for irrigation, drainage, and general purposes ; but they are frequently neither efficient nor capable of dealing with large quantities of water.   Among them may be mentioned the "tympanum-wheel," "Persian wheel," "scoop-wheel" and "chain-pump."

Fig. 23.

The tympanum-wheel, *Fig. 23*, raises water by the revolution of its curved radial vanes (which are closed at the sides by a pair of flat plates), from the lower part of its periphery up to the hollow axis on which it revolves. The water, rising as it follows the diminishing curvature of the vanes to the centre of the wheel, passes through an aperture in the axis corresponding successively with holes in the central part of the revolving case, and is thence delivered.   This machine is frequently arranged to be driven by the flow of a stream from which it raises water, and then forms a convenient and durable, if inefficient, apparatus.

The Persian wheel, *noria* or "bucket-wheel," made up to 30 feet in diameter, carries a series of buckets, or earthenware pots, upon the circumference, so arranged as to lift water from the trough into which they dip at the bottom of the wheel, and, by various devices, to empty it at or near the top. The buckets are now commonly carried on an endless band, as was occasionally the case in ancient examples of this apparatus.

Scoop-wheels, which may be described as breast water-wheels caused to revolve backwards, are found, in the Fens of England and elsewhere, convenient for adapting wind-power to lift water a limited height. A wheel 50 feet in dia-meter, which is about the largest size made, will raise water about 15 feet, with a peripheral speed of 6 feet per second. They are, however, wasteful in action—slip, and the leakage between the wheel and its side-walls causing a serious loss—and are nowadays often dis-placed in favour of centrifugal pumps driven by steam-power.

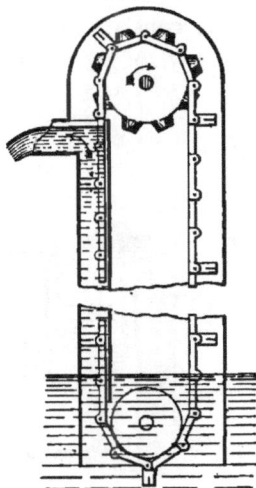

Fig. 24.

The chain-pump, *Fig. 24*, con-sists of two parallel shoots, with wheels at the top and bottom, round which an endless chain passes. To the chain are attached, at intervals of 3 or 4 feet, flat plates which fit closely the shoot up which the water is raised. The power is applied to the upper wheel, the motion of the chain causing the plates to lift a continuous stream of water up the rising shoot. The chief objection to this apparatus consists in the number of its joints and working parts, but it is useful to pump gritty water or liquid mud. Although not economical in action for any but very low lifts, owing to

the dead-weight of the apparatus to be moved, it is capable
of working with a lift of 50 or 60 feet.

157. Among force-pumps may be noticed the "spiral
pump," the "pulsometer" and the "hydraulic ram."

The spiral pump, invented by a Zurich carpenter in the
middle of the 18th century, is an ingenious apparatus
that has received but little attention in this country. Its
employment in irrigation works abroad, and the peculiar
simplicity of its action, may warrant us in giving a short
description of it.

It consists in its simplest, though least effective, form,
Fig. 25, of a cylinder or drum, round which is coiled
a pipe open at one
end, and terminating
in a hollow journal at
the other end. Into
this journal passes,
through a stuffing-
box, a pipe that forms
a rising-main, which is
furnished with a foot-
valve. The drum is
partly immersed in a
stream of water, in
which it is caused to
revolve by means of
paddles fixed to one
end of its axis. As it
revolves, charges of
air and water are al-
ternately taken in by
the open end of the pipe, which is trumpet-mouthed,
and travel along the spiral through the hollow journal into
the raising-main, up which the air at once escapes. The

Fig. 25.

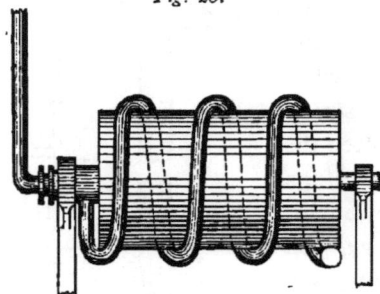

Fig. 26.

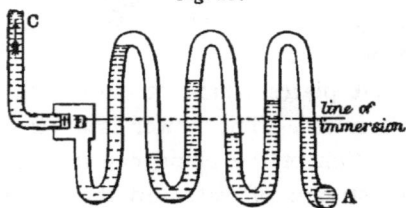

*line of
immersion*

A, inlet ; B, foot-valve ; C, rising-main.

height to which the water can be forced by this apparatus depends upon the diameter of the drum and the number of turns in the coil of pipe. As the water advances along the spiral, its level on the forward side of the hollows is more and more depressed as the air becomes compressed to a smaller volume in its advance towards the journal, as shown in the diagram, *Fig. 26.* The pressure in the latter is consequently the sum of the differences of level of the water in each successive hollow of the spiral.

The delivery of the pump varies with the relative quantities of air and water admitted at each revolution. Owing to the volume of each charge of air diminishing in its progress through the coil, the quantity of water increases from coil to coil, and this increase, if considerable, as in the case of a machine working near to its maximum lifting capacity, is apt to cause momentary derangement of the action; although the foot-valve in the rising-main effectually prevents danger from the fall of the column of water in the latter. This difficulty is obviated when the drum round which the pipe is coiled is made conical, so that the quantity of water taken in at the mouth is, with the air, sufficient to fill the smaller coils, notwithstanding the decreasing volume of the air.

The condition of maximum efficiency then occurs when the drum is half immersed in the water, so that equal volumes of air and water are taken into the pump, and when the number of coils is such that the lower surface of the water on the forward side of the last hollow is level with the bottom of the coil, and its upper surface is level with the top of the coil, as illustrated in *Fig. 26.*

The lift H of which the machine is capable, is then found approximately from the expression

$$H = n\,r,$$

where r is the radius at the small end, and n the number of coils in the spiral.

To find r :

Let R be the radius of the inlet end, and L the atmospheric pressure expressed as head of water,—

Since the difference of the water-levels in the last coil is $2r$ and in the first one o, the volume of air in the last coil is

$$2\pi r - \pi R \quad \text{or} \quad \pi (2r - R).$$

And since the volume of the air in the first coil is πR, and the pressure at the inlet is L, and at the outlet $H + L$, it follows that

$$\frac{\pi (2r - R)}{\pi R} = \frac{L}{H + L},$$

or

$$r = \frac{R}{2} \left(1 + \frac{L}{H + L} \right).$$

The friction of the apparatus is small, as the drum can be floated in the water, so as to bear lightly on its journals; but the intermittent discharge of air, accompanied by a surging of the water in the rising-main, is a disadvantage, and is probably mainly responsible for the want of attention experienced by this type of pump. The loss of efficiency in the above forms of pumps is chiefly due to friction and to slip.

158. For temporary purposes, the pulsometer is a most useful pump, being driven by steam without the intervention of mechanism and capable of working suspended from a chain or rope in almost any situation. It consists of two pear-shaped vessels, inclined towards each other, and meeting at the top in a chamber in which a spherical ball in unstable equilibrium closes the neck of either one or the other vessel. Above the level of this ball-valve the steam-pipe communicates with the chamber. In the

bottom of each vessel is a suction-valve, communicating with a common suction-pipe to which an air-vessel is connected. Valves from both chambers open also into a common delivery-pipe.

The action is as follows : the vessels are at starting filled with water through plug-holes ; steam is then admitted into the valve-chamber, and passes into the vessel left open by the ball, forcing the water in it down to the level of the delivery-valve, when a little steam blows violently through the water into the delivery-pipe. Instantaneous condensation takes place, and a vacuum is consequently formed in the vessel. This causes the ball-valve at the top to be forced over, shutting off the steam-connection with this vessel, when condensation is completed, water meanwhile entering the vessel through the suction-valve. The other vessel is now open to the steam-supply, and exactly the same cycle of operations takes place in it, the action proceeding in the two vessels alternately, as long as steam is supplied to them. The consumption of steam in this pump may be greatly reduced (though the quantity of water pumped with a pulsometer of given size is also reduced) by the addition of a "Grel" valve, which causes a secondary cut-off and produces expansive use of the steam.

159. The hydraulic ram employs the momentum of some considerable body of water, having a slight fall, to elevate a small quantity of it to a comparatively great height. This apparatus is shown diagrammatically in *Fig. 27.* It consists of a strong main pipe laid at such an inclination as is available, and connected at its upper extremity with the stream or reservoir whose water is to be utilised. Near the lower end of the main pipe, it is provided with a waste-valve of large area, opening inwards, and at its extremity is a small delivery-valve opening

outwards and communicating with an air-vessel, from the bottom of which the delivery-pipe ascends to the tank or

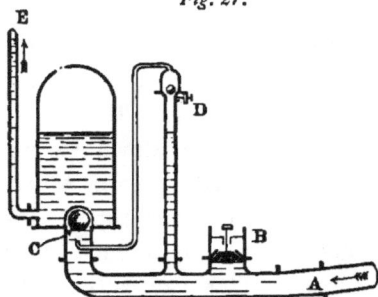

Fig. 27.

A, main supply-pipe; B, waste-valve; C, delivery-valve; D, snifting-valve; E, delivery-pipe.

cistern into which the water is to be pumped. Between the waste- and the delivery-valve is a small branch pipe furnished with a reflux-valve at the top, and, on the lower side of this, with a "snifting-valve," or fine orifice the area of which can be varied by a screw-plug. Connection is made by a small tube between the upper side of this reflux-valve and the lower side of the delivery-valve.

When the ram is in action, the main pipe is supplied with water, which issues freely through the open waste-valve until the velocity of its motion is sufficient to raise and close this valve. The body of water then in motion in the pipe being quickly arrested, its impulse causes the delivery-valve to open and a small quantity of water to be forced into the air-vessel. Simultaneously, the air in the branch pipe is, notwithstanding a small loss through the snifting-valve, so compressed as to raise the reflux valve momentarily—when a discharge of air takes place through the delivery-valve into the air-vessel. The effect of the impulse or "ram" of the water having subsided with a slight rebound, due to the elastic nature of the resistances that have contributed to check its motion, the pressure in the main pipe falls, the waste-valve opens again, and the water falling in the small branch pipe sucks in a fresh charge of air through the snifting-valve.

The object of the air-pump arrangement afforded by the branch pipe and the reflux- and snifting-valve, is to automatically replenish the main air-vessel—an important feature of the apparatus, especially with high lifts. It will be readily understood that a machine acting by a series of violent impulses must, in any case, be somewhat limited in its application ; and this form of pump (the illustration is to be regarded as diagrammatical only) is, where used, restricted to comparatively small supplies.

By the introduction of a cylinder and piston, with suitable valves, between the working-barrel or main pipe of the ram and the delivery pipe, a large stream of impure water may be utilised to pump a small quantity of pure water from an adjacent source.

The efficiency of hydraulic rams is affected principally by the water necessarily lost through the waste-valves, by friction in both the supply- and the delivery-pipes, by the resistance at the valves, and by the distention of the apparatus under the high pressures that occur when the waste-valves close. If H is the height of the upper end of the supply-pipe above the waste-valve, h the height at which the water is delivered, Q the rate of supply and q that of delivery ; and if H_0, h_0 and h_1 denote the head lost by the resistances in the supply-pipe, delivery-pipe and at the waste-valve, respectively,

$$Q (H - H_0) = q (h + h_0) + (Q - q) h_1,$$

or

$$Q (H - H_0 - h_1) = q (h + h_0 - h_1) ;$$

that is,

$$\frac{q}{Q} = \frac{H - H_0 - h_1}{h + h_0 - h_1}.$$

In practice, h_1 may be expected to be about $\dfrac{H}{4}$.

160. The hydraulic-pressure pump, *Fig. 28*, possesses many of the advantages of the hydraulic ram, and is free from the objectionably violent action of the latter. By its use, a large quantity of water under a small head, flowing slowly through a supply-pipe, may be caused to steadily raise a small portion of the water to a high level. A vertical cylinder of large diameter, open at the top, supports on radial brackets a smaller cylinder surmounted by a valve opening upwards into an air-vessel. The delivery-pipe leads from the bottom of the air-vessel in the usual way. Through a piston that works in the large cylinder, passes a hollow piston-rod capable of a small range of motion, fitting a hollow plunger attached to the main piston, and furnished at the top with a ball-valve opening upwards. This hollow plunger works in the small upper cylinder. The inlet- and the outlet-valve of the main cylinder, situated under its centre, are opened and closed by disks on the piston-rod. A lateral orifice in the latter affords communication between the water under pressure in the inlet-pipe and the hollow plunger.

Fig. 28.

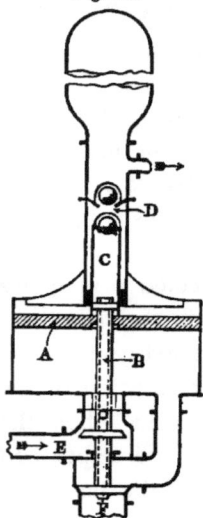

A, main piston;
B, hollow piston-rod;
C, hollow plunger;
D, upper cylinder;
E, supply-pipe;
F, outlet-valve.

This orifice and the main inlet- or supply-pipe being open, the piston is raised, and with it the plunger, forcing water with considerable power out of the small upper cylinder into the air-vessel and delivery-main, and drawing water through the piston-rod into the hollow plunger. Near the top of its stroke the piston engages with a collar on the piston-rod, raising the latter so as to close the inlet-valve, open the outlet-valve, and simultaneously to cover the orifice behind a fixed collar through which the piston-rod slides. The main piston then sinks as the water beneath it escapes through the outlet-valve, and as it descends the piston-rod displaces water from inside the plunger into the cylinder above, ready for the next stroke of the piston. By duplicating the cylinders, the pump may be caused to deliver water both steadily and continuously.

161. A review of the apparatus employed for raising water almost necessarily involves reference to steam-engines and other prime movers. Although mechanical contrivances in detail are outside the scope of this treatise, the general conditions of efficiency of certain steam-engines employed for pumping, and the methods according to which it is estimated, will be discussed briefly. In order to render the matter intelligible, the general principles of action of the several types of engine will be first sketched; though for a clear comprehension of it, their structure should form a subject of special study.

The expansive working of steam-engines that are used to drive pumping-machinery, introduced for economical reasons, obviously gives rise to the condition that the power applied by the piston at each point of its stroke generally differs considerably from that required at the same instant to work the pumps, assuming uniformity in the delivery of the latter (§ 146).

There are three methods commonly practised for overcoming this, and equalising the effort exerted upon the pump to the resistance of the latter :—

(1) By the use of an oscillating beam, or a rotating fly-wheel, which, by increased momentum during the portion of the stroke at which the power developed exceeds that

required by the pumps, accumulate energy that is restored when the power exerted by the piston falls below the average, as in the Cornish or the ordinary rotative pumping-engines. (2) By so arranging the steam-cylinders, and linking the pumps together, that, with the momentum of the moving parts, the effort of the pistons equals at each moment the resistance of the pumps, as in the Davey engine. (3) By causing the excess of the effort of the pistons in the early part of the stroke either to compress air, or to pump water into an accumulator, stand-pipe, or pressure-main. The energy thus stored supplements that of the steam during the latter part of the stroke of the piston, as in the Worthington type of engine.

162. Of single-cylinder engines, the old Cornish beam-engine, in which steam is admitted on one side of the piston only, is still the most economical for the range of expansion employed in it. Cornish engines are of two kinds : beam and direct-acting. In the beam-engine the steam is admitted above the piston. The piston-rod is connected with one end of a stiff iron or steel beam of considerable weight, the pump-rods being attached to the other end. The weight of the rods is sufficient to perform the "out-doors" or up-stroke of the piston. No fly-wheel is employed, and the engine cannot be governed in the ordinary way. The steam-, equilibrium- and exhaust-valves are double-beat, worked by tappets. The rate of working is governed by a cataract, which regulates the opening of the valves.*

163. The direct-acting, or "Bull," engine is constructed so that the pump-rods form a prolongation of the piston-rod, and the steam is admitted below instead of above the piston as in the beam-engine. The arrangement of both

---

* For a description of the cataract, *vide* Downing, 'Elements of Prac.ical Hydraulics,' p. 169—London, 1875.

engines is such that the steam side of the piston is never open to the exhaust; consequently, that part of the cylinder and piston is but slightly cooled at each stroke. This circumstance accounts for the fact already mentioned, that, for the same range of expansion, a Cornish engine is more economical than any other single-cylinder type. The "duty" of a good Cornish engine and pump may be expected to reach 90,000,000 foot-pounds for one hundred-weight of good steam-coal consumed in the boiler-grates.*

164. Compound engines are either of the "Woolf" type, in which the exhaust-steam from the high-pressure passes directly into the low-pressure cylinder, or of the "receiver" type, in which the high-pressure cylinder is exhausted into an intermediate receiver before being used in the low-pressure cylinder.

165. In the rotative beam-engine, which is furnished with a fly-wheel driven by a connecting-rod directly from one end of the beam to the other end of which the connecting-rods of the two pistons are attached, the points of attachment of the latter, the main pump or pumps, the air-pump and the circulating pump to the beam may vary considerably. An object that must be kept in view, is the reduction of the stresses in the beam and of the resultant pressure on the gudgeons that carry it, to a minimum. By prolonging the piston-rod of either the high-pressure or the low-pressure cylinder through the bottom cover to work a pump beneath, the gudgeons may be relieved of considerable pressure, and the stresses in the beam may be reduced. A similar result may be effected, though in a less degree, by connecting an auxiliary pump with the cylinder side of the beam. An "entablature," or horizontal iron frame

---

* The duty of pumping-engines is expressed by the product of the weight in pounds of the water raised by the pumps, into the height in feet to which it is lifted, resulting from the consumption of a given quantity of coal in the grates of the boilers that supply the steam (§ 173).

supporting the beam, is either built into the walls of the house, or is carried by columns. In the latter case, in addition to the columns, vertical A-frames are introduced beneath the gudgeons to resist the longitudinal thrust. This arrangement, whilst increasing the cost of the engines, reduces somewhat that of the engine-house. The cylinders are bolted directly to the bed-plate, which may span the well.

166. The beam-engine permits of a long stroke and a high piston-speed in the steam-cylinders, whilst at the same time a shorter stroke and lower speed may be given to the pumps ; and in pumping from deep wells it is a most serviceable form of engine. It is easily balanced, and affords facilities for working more than one pump from the beam. In common with the type of engine to be next described, it possesses the advantage that all the bearings can be kept tight by adjusting the brasses in a vertical direction only. The cylinders and pistons tend in working to preserve their circular shape, lubrication acts uniformly round their circumference, and they are easily kept free from internal moisture. This form of engine may conveniently work up to 25 revolutions per minute, according to the length of the pump-rods and the size of the pump-valves.

167. The vertical direct-acting rotative engine may frequently be well employed when only one main pump is required. The cylinders, compound or triple, are placed vertically over one another, their common piston-rod being prolonged downwards to work the pump, transmitting the steam-pressure directly to the plunger and, by a connecting-rod, to the fly-wheel. The number of working parts and the weight are thus considerably diminished below those of the beam-engine ; and it may be run at a higher speed than the latter—to 30 revolutions per minute. The engine-

frame is independent of the house, and the area required for it is very small. Up to 100 horse-power, there is no difficulty in equalising the effort to the work by means of a fly-wheel; but in larger sizes, a small beam may be usefully added, from which the auxiliary pumps are conveniently driven.

168. When the depth of suction admits of the employment of a horizontal engine having the pump in line with it, the horizontal rotative engine forms the cheapest and simplest machine of its class. If, however, the use of a horizontal engine involves an approach to 30 feet depth of suction, it is often unsuitable for the purpose, as this condition is disadvantageous to the easy working of large engines. A convenient arrangement is afforded by each piston actuating a separate crank on the same shaft, having its own main pump in line with it. An air-pump for the condenser, and in certain cases a circulating-pump, are worked from the tail-rods of the main pumps. An expansion-valve, adjustable by hand, on the low-pressure cylinder, enables the work done upon both pumps to be equalised, a second expansion-valve being provided for the high-pressure cylinder, and regulated by a governor.

In raising water from deep wells, vertical pumps are frequently worked from horizontal engines by means of rocking-levers, at the expense, of course, of some additional loss of power in friction. The horizontal engine requires little head-room, but considerable floor-space.

169. The Davey engine is made in various forms. Its objects are to use steam expansively whilst doing a constant amount of work, and, by differential valve-gearing, to adjust its action to sudden variations of load, should such occur.

The first object is attained as in Cornish engines: (1) by taking advantage of the momentum of the pump-rods

M

or, if these are short, that of a moving weight or an oscillating disk ; or (2) by suitably arranging two sets of vertical steam-cylinders, and coupling together, by a rocking-frame, the pumps working underneath them.

The former arrangement is chiefly used in pumping-engines that drain mines, where the pump-rods possess considerable inertia. In the latter arrangement, the rocking-frame forms a varying lever by means of which the resistance of the pumps is caused to admit of a higher degree of expansion than would be possible in employing the inertia of the pump-plungers alone. The mechanical principles alluded to are embodied in simple, compound and triple engines. In the compound engines, the cylinders are placed horizontally, vertically or inclined to one another, as may be required to suit the varying conditions met with in practice.

In the second form of this engine, the high- and low-pressure cylinders form one chamber, the piston of the former being connected with that of the latter, which is vertically above it, by a trunk of its own diameter. The steam-cylinders and pumps are in duplicate, the latter being situated below the former and actuated directly by the piston-rods, the pump-plungers being connected together by means of a rocking-frame. Both pistons being at the beginning of their stroke simultaneously, steam is admitted under the high-pressure piston, and is, after cut-off, expanded into the annular space around the trunk in the lower part of the other low-pressure cylinder. The up-stroke being completed, the steam is transferred to the upper side of the low-pressure piston, when the down-stroke is effected by a third expansion. The expansion-valves of the steam-cylinders are connected by levers with the rocking-frame, and the main valves are worked from the piston of an auxiliary steam-cylinder, uniformity of motion being ensured by a cataract.

The effect of gearing the valves in this manner is that if the engines suddenly lose their load, and the pistons begin to race, the motion of the expansion-valves soon attains the same phase as that communicated to the main valves from the uniform movement of the auxiliary piston. The steam is consequently cut off earlier in the stroke, and the piston-speed is reduced to its normal value. If, on the other hand, an additional load is suddenly thrown upon the engines, the pistons commence moving more slowly, and the motion communicated from the pistons to the expansion-valves takes longer to attain the same phase as that of the main valves. Consequently the cut-off is delayed, and the piston-speed is thus restored to its former amount. This form of regulation, affecting the cut-off in both the high- and low-pressure cylinders, is more reliable and rapid in action than that of a governor regulating the point of cut-off in the high-pressure cylinder only.

By a recent arrangement of the differential gear, in addition to cutting off steam from both cylinders when the engines race, the exhaust-valve from the high-pressure cylinder is closed.

The Davey engine combines the advantage of a high rate of expansion with perfect safety in the event of a pump breaking down, or of a burst occurring in the delivery-main.

170. The Worthington direct-acting engine possesses the great advantage of working with a ratio of expansion as high as that of any type of rotative engine, at the same time exerting a uniform pressure upon the pump-plunger.

It consists of two similar engines side by side, the motion of one engine actuating the valves of the other. Expansion-valves on both the high- and low-pressure cylinders are actuated by the engine to which they belong. The main valves are of the Corliss type. Two of these on the high-pressure cylinders serve for both inlet and exhaust. The low-pressure cylinders are each provided with two inlet- and two exhaust-valves. The ratio of expansion is variable by hand.

The pumps are bolted to the bed-plate, but the steam-cylinders are left free to move lengthwise as change of temperature may require. The former are worked by the piston-rods of the engines, and each pump consists of a chamber divided by a transverse partition, in the middle of which is a collar, through which a solid plunger works. Numerous small inlet- and outlet-valves are formed in the chamber-wall, on both sides of the partition.

At the end of each stroke a pause occurs, to allow the pump-valves to close before the commencement of the next stroke, by which action "slip" through them is prevented. There is no fly-wheel, and the pressure exerted by the engines is rendered uniform by means of compensating-cylinders in the following manner. The high- and low-pressure cylinders are arranged for such an expansion that the excess of their combined pressure in the first half of the stroke above the mean pressure, is equal to the defect in the second half of the stroke below it. Two oscillating cylinders are placed, facing one another, one on each side of the piston-rod, with their trunnions at right-angles to it and attached to the main framing. The piston-rods of these cylinders are connected with a sleeve on the main piston-rod. When the latter is at half-stroke, the compensating (oscillating) cylinders are at right-angles to it. During the first half of the stroke of the engines, water is forced into an accumulator or against air-pressure by the pistons of these cylinders, and the energy thus stored is given back in the second half of the stroke. The size of the compensating-cylinders, and the pressure of the water in them, is arranged so that the work done in the first half of the stroke, and returned in the second half, is equal to the excess and defect respectively of the combined work of the high- and low-pressure cylinders above and below the mean, in the first and second halves of the stroke of the main piston. For waterworks purposes, the pressure in the compensating cylinders may be adjusted by suitable devices to the varying head of water against which the engines are pumping.

171. All the engines mentioned are worked with either jet- or surface-condensers. It is now a common practice in the latter case to allow water from the main pump to pass through the condenser. The temperature of the water is but slightly raised thereby, and a circulating-pump is dispensed with.

172. The generation of steam for pumping-engines is a subject with which we are only incidentally concerned, because, in the remarks which we propose to offer on the testing of such engines, their efficiency is measured in relation to the steam consumed by them, and is thus separated from the question of the efficiency of steam-generators. If any doubt exists as to the wisdom of this course, we commend to the notice of the reader Mr. J. G. Mair-Rumley's excellent paper on " The Independent Testing of Steam-Engines." *

173. The efficiency of a steam-engine may be defined as the ratio which the power developed in its cylinders, expressed in thermal units, † bears to the heat supplied to the engine.

To distinguish the efficiency of the engine from that of the boiler which supplies it with steam, the number of thermal units consumed by the engine must be measured,

---

* Minutes of Proceedings Inst. C.E., vol. lxx. p. 313.
† It is still customary in this country to use the Fahrenheit scale of temperature in this particular branch of scientific inquiry.

as well as the fuel consumed in the boiler-grates, and compared with the indicated work of the engine.

As the total heat contained in a pound of dry steam does not vary greatly through the ordinary range of pressures, the number of pounds of dry steam required to develop one horse-power in the cylinders may be regarded as an approximate measure for the comparison of the efficiencies of engines.*

To form this comparison, indicator-diagrams, showing the power developed in the cylinders, are taken at intervals of ¼ hour during the period of the trial, the quantity of feed-water and its temperature are measured—the boiler-pressure being also observed and the fuel burnt being weighed. These observations afford the necessary data for calculating the number of thermal units communicated to the engine for each horse-power developed in the cylinders, as well as the evaporative effect of the boiler per pound of coal burnt—a method of engine testing which was initiated by Mr. G. A. Hirn.†

174. Where possible, the amount of heat rejected from the engine should be ascertained as a check upon the heat supplied to it by the boiler, for the heat supplied to an engine equals the work done expressed in thermal units, together with the heat rejected.

Where boilers supply steam to more engines than that under test, or for other purposes, and such supplies cannot be separated, the measurement of the heat rejected added

---

* It need scarcely be mentioned that the mechanical efficiency of an engine is the inverse ratio which its cylinder-power bears to the power developed by its driving-shaft or other motive agent; and that the efficiency of a pump driven by it, is the inverse ratio borne by this latter quantity to the work done, as measured by the product of the water raised into the actual lift of the pump.

† Hirn, ' Exposition Analytique et Expérimentale de la Théorie Mécanique de la Chaleur '—Paris, 1862, and later editions.

to the indicated work affords the only means of finding the
total heat supplied to the engine.

175. The point of cut-off, and the steam-pressures at
admission to and exhaust from the cylinders are carefully
measured and averaged whilst the indicator-diagrams are
taken. In a paper read by Mr. A. W. Brightmore in
1885,* it was shown that the effect of the use of a flexible
cord to give motion to the drum of the indicator is to
distort the diagrams. This effect may be avoided by the
use of a steel wire. To ensure accurate results, the springs
of the indicator should be carefully tested, both for pressure
and vacuum, before being used. The areas of indicator
diagrams are best measured by a planimeter, and the
average of all those obtained during the trial is taken.

176. The feed-water to the boilers is measured into
the tanks from which the boilers are supplied, and the
quantity of water in the boiler is, at the conclusion of the
trial, left the same as it was at the beginning. When
convenient, the water from the steam-main is collected in
buckets, in order to give an idea of the amount of priming
that takes place ; but it is generally arranged for this water
to drain back to the boiler.

An estimation of the heat rejected in the exhaust is
more easily arrived at by ascertaining the discharge
from the air-pump than by measuring the amount of the
injection-water. The former quantity is therefore usually
measured, and is best done by causing the water to pass
through one or more orifices, any variation in the head
above these being recorded. The calculations applicable
to the discharge of such orifices (§ 80) should be checked
by actual experiment if very accurate results are desired ;
as slight differences in their mechanical finish may exert a
considerable effect upon their discharge.

* Minutes of Proceedings Inst. C.E., vol. lxxxiii. p. 20.

The weight of the total discharge of water thus found
is equal to that of the steam passing through the cylinders,
together with the injection-water.

177. The heat rejected is equal to the product of the
weight in pounds of steam passing through the cylinders
into the number of thermal units due to the temperature
of the air-pump discharge, and the weight of the injection-
water multiplied by the number of thermal units due to
its rise of temperature in the condenser. As the specific
heat does not vary much between these temperatures, the
differences in the latter may be taken as the multipliers.

178. Since the air-pump abstracts air which escapes
laden with moisture at the temperature of the hot-well, a
certain amount of heat is lost in this manner and is not
registered ; however, in carefully conducted experiments,
the measurement of the heat rejected should coincide with
the difference between the heat supplied and the work
done, within at most 5 per cent. The heat rejected in the
exhaust cannot be calculated in the above manner if a
surface-condenser is used.

The steam that condenses in the jackets is collected
in suitable vessels, the drain-cocks being open as far as
possible without letting the steam blow through. The
heat rejected in this manner equals the weight of water
collected, less the priming-water in the jacket-steam,
multiplied by the number of thermal units corresponding
with its temperature.

The radiation from the cylinders is found by turning
steam into the jackets and cylinders, when the engines are
standing, and measuring the condensation in the jackets
during a period of several hours. The weight of this
water, multiplied by the latent heat of evaporation at the
temperature of the steam, equals the heat lost by radiation
during this time.

179. It thus appears that the heat supplied to an

engine is equal to the weight in pounds of dry steam supplied to it, multiplied by the number of thermal units added to one pound of dry steam between 32° F. and the temperature corresponding with the boiler-pressure. This is also equal to the work done in the cylinders of the engine expressed in thermal units, together with the heat rejected into the condenser above the temperature 32° F., the loss of heat due to condensation in the steam-jackets and the loss by radiation.

From these relations, the objects to be sought in an investigation of the efficiency of a pumping-engine are at once apparent ; and the foregoing outline of the methods adopted to determine the several elements of the question may, whilst it does not pretend to be exhaustive, serve to place before the reader a concise summary of the operations involved in it.

180. The mechanical efficiency of well-constructed engines of the types discussed in the foregoing sections ranges as high as 90 per cent., and it may be expected (§ 151) that an efficiency of 95 per cent. should be obtained in the pumps driven by them. The "duty" of the entire pumping-plant frequently attains 120,000,000 foot-pounds to 130,000,000 foot-pounds (water lifted) for each hundred-weight of good steam-coal consumed in the boiler-grates.

181. Besides steam, pumping-engines are, under special circumstances, driven by power derived from water under pressure or in motion under gravity, from the combustion of gas or oil, and from the wind.

As explained in § 65, Bernoulli's theorem

$$\frac{v^2}{2g} + \frac{p}{w} + h = H$$

represents the total energy along any stream-line, and the various forms of hydraulic motors utilise the energy of water as expressed by one or more terms of this equation.

182. When water available for motive purposes exists under a head of more than 200 feet, water-pressure engines are generally found to be suitable forms of motor, the energy being utilised as expressed by the second term of the equation ; as is also the case in the hydraulic-pressure pump (§ 160).

Under more moderate heads, turbines afford an economical means of utilising the energy of the water ; or, if the head does not exceed 50 feet, overshot water-wheels may be employed with advantage. Re-action wheels utilise the energy of the water in the form expressed by the first term of the equation, as do turbines if of the "impulse" type (which are most applicable when the supply of water is variable) ; and in that given by the first and second terms, if of the re-action type. When the head does not exceed 6 feet, the best water-motor is probably the undershot water-wheel, in which the energy of the water is utilised in the forms expressed by the first and third terms of the equation—chiefly the former. The hydraulic ram, which is at once a pump and a motor, utilises the energy of water as expressed by the first term of the equation, after having converted it into the form expressed by the second term.

Unfortunately, it happens that a satisfactory source of water-power is seldom available in the positions required for pumping-works, and we do not propose to further discuss the theory or application of this form of motive power.

Gas and other internal-combustion engines are also of such inconsiderable importance in this connection, as hardly to merit special treatment.

183. Motive power derived from the wind is extensively used in pumping for drainage and irrigation purposes, and may sometimes be usefully employed in village water-

supply works. In pumping for drainage purposes, owing to the sub-soil forming a natural reservoir, wind-power is usefully employed to lower the water-level when it happens to be available, but for the remainder of the time the surface of saturation gradually rises in the ground. When employed in connection with the water-supply of small districts, unless supplemented by steam or other power, it requires storage-tanks to ensure continuance of the supply when there is no wind ; and it must generally be regarded as an unreliable, if economical, source of power for such an important function as that of raising water for domestic use.

CHAPTER IV.

184. IN storing water for ordinary purposes of supply it generally happens that two somewhat opposite conditions have to be satisfied. On the one hand, it is frequently required that a fluctuating quantity of water yielded by the inconstant sources in nature, shall be so averaged as to satisfy the more regular requirements of civil life ; whilst, on the other hand, a steady and uniform supply has to be adapted to meet the periodic variations of domestic and industrial consumption.

In waterworks that are dependent directly upon rainfall collected on comparatively small catchment-areas, provision must be made to afford supplies that are to a large extent independent of the variable yield of water which is the result of meteorological phenomena—cyclical or erratic ; whereas, in those which draw supplies from great rivers or lakes, or from the extensive water-bearing formations of the earth's crust, such provision has not to be specially contemplated, because it is afforded naturally. But either class of works is seldom adapted to effect its objects without some arrangement by which the extreme diurnal variation of consumption may be met by a comparatively constant delivery of water at or near the place of its distribution.

185. The first-mentioned requirement is satisfied by the formation of " impounding-reservoirs," the office of which is to gather the irregular natural yield of surface-water, in

order that it may be supplied at a uniform rate ; the second
is met by the construction of "service-reservoirs," tanks
and cisterns, from which water is distributed as required by
the .hourly demands of consumers. Another important
function of storage-reservoirs is to render. possible a constant supply of ·compensation-water (§ 34) to the streams
affected by the waterworks to whose service they have been
converted. And it is to be remarked that such streams
derive no inconsiderable advantage from the mitigation of
their floods, brought about by the impounding of the water
required to fill and raise the level of the broad expanse of
the reservoirs—an action that is operative even when the
latter are brim-full.

An efficient and economical reservoir must be capable
of maintaining a fair balance between the demand for and
supply of water during any given period of time, and .yet
must not exceed the minimum size necessary to effect this
object, if such additional size entails increased cost in construction. The capacity of reservoirs is therefore a vital
and elementary question of every scheme of water-supply.
Where the requisite data are available, it becomes a mere
mathematical matter ; although it frequently happens that
the experience and judgment of the engineer must be
exercised in the preparation or selection of such data, to an
extent that can only be fully realised by those who have
engaged in the operation.

186. The absolute necessity for an impounding-reservoir,
as part of a waterworks, arises when the rate of consumption
during some part of the year is higher than the natural
yield of the source of supply. The condition to be fulfilled
is that the surplus water yielded during periods of excessive
rain or flood shall be retained and utilised to supply such
natural deficiency in times of drought. How much of this
surplus is to be thus retained depends entirely upon the

exigencies of each particular case. Hence the proper capacity for the reservoir is governed by the relation that subsists between the natural yield of the source and the rate of consumption of the water. The latter of these two elements is given by the specified conditions of the water supply in question—whether estimated upon the basis of population and domestic requirements, or upon that of power required, or compensation, or lockage in the case of canal-supply, or other disposal of the water for industrial purposes, as the case may be.

The former element is directly ascertained by gauging the flow of water off the area or in the river under consideration, or is inferred approximately from the rainfall, according to the methods explained in §§ 61, 62.

187. The office of the impounding-reservoir being, broadly speaking, to store up water in times of plenty and to dole it out in seasons of deficiency, it might appear that the requirements of any particular case would be met by making the reservoir of capacity sufficient to ensure the maintenance of the usual daily supply during the longest drought that experience might lead us to apprehend. When, however, it is considered that prolonged periods of alternate drought and flood may occur, in which the successive spells of drought, although not individually of remarkable duration, may, in the aggregate, far outweigh the shorter alternate spells of wet weather, it becomes obvious that any single term of dry weather does not form a sufficient basis to proceed upon in designing the capacity of the reservoir-accommodation. Further, as either the whole, or only a part, of the absolute yield of the catchment-area may be wanted to maintain the supply, it is evident that any rules for guidance in this matter must, in order to be generally applicable, take into account, not only the physical conditions that affect the amount and

periodic variations of the yield of water, but also the conditions under which the supply is to be drawn.*

188. Empirical formulæ have been suggested for determining the storage-capacity required for a given supply from a knowledge of the mean rainfall of three consecutive dry years, without reference to the rate of supply provided or the actual fluctuations of the yield—though probably on the assumption that all the available water except that of great floods should be stored. The application of such formulæ must, however, be very limited, or must involve re-determination of the constants in order to suit variation of either meteorological conditions, or the geological nature of the district under consideration, or the rate of supply.

E. g. an approximate formula for use in this country is

$$C = \frac{1000}{\sqrt{mean\ rainfall\ of\ 3\ dry\ years,\ in\ inches}}$$

where C is the number of days' storage required. And, in this connection, it may be noticed that the actual storage provided in the waterworks of Liverpool, Manchester, Dublin and Edinburgh differs from the figures deduced by this rule between 10 and 30 per cent.

189. The great expense incurred in the formation of reservoirs, and the important nature of their duties, renders it imperative alike in the interests of economy and sanitary efficiency that their storage-capacity should be made the subject of careful inquiry as a preliminary step towards their design. The most satisfactory data upon which to proceed is a long series of measurements of the yield of the source in question. In the absence of such data, recourse may be had to an estimation of the yield during three consecutive dry years, one of which should be a phenomenally dry one. In forming this estimate from rainfall statistics,

* Compare Rankine, 'Civil Engineering,' 10th edition, p. 700.

it is highly necessary to consider the geological character of the district, and how far it may of itself obviate the necessity for artificial storage, by equalising the dry-weather and wet-weather flow of the streams. It would be tedious to enter into a disquisition upon all the various observations which, under different circumstances, would lend aid to this inquiry. Much follows directly from the application of the principles treated in Chapter II. The remainder must be left to the practical knowledge gained by the engineer in other catchment-areas.

190. Could both the yield from the source, and the desired supply, be expressed as periodic functions of the time, the question would admit of easy algebraical treatment. The generally erratic character of the yield, however, renders it more convenient to deal with it by graphic analysis. Several processes are in use for this purpose.* The following solution of the question was communicated in 1891 to the Liverpool Engineering Society : †—

Having ascertained the monthly increments of both yield and supply during the period under examination— expressed either as available rainfall in inches, or as an actual quantity of water measured in cubic feet—from the addition of these quantities form two series, giving respectively the total yield and the total desired supply up to the end of every month, from the commencement of the period.

Plot these series of figures as ordinates upon a diagram, *Fig. 29*, using the corresponding time, in months, as abscissæ plotted along OX, so that the ordinate at any epoch measures the total yield, or supply, as the case may be, during the period then terminating.

Through the upper extremities of the two sets of ordi-

* Minutes of Proceedings Inst. C.E., vol. lxxi. p. 270.
† Trans. Liverpool Eng. Soc., vol. xiii. p. 53.

nates, draw two curves OA, OB, which may be termed
curves of "yield" and "supply," respectively.

From the construction, it is evident that the inclination
of either curve to the horizontal axis, at any point, measures

Fig. 29.

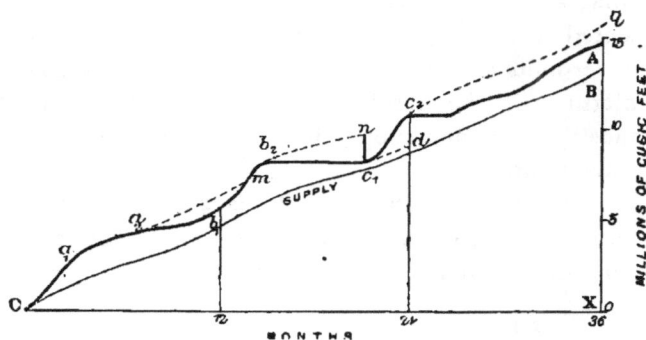

the rate of increase of the quantity symbolised by the curve
considered. Where, as at $a_1 a_2$, the "yield" curve runs
exactly parallel to the "supply" curve, the yield from the
catchment-area just equals the supply.

Where, as at b_1 b_2, c_1 c_2, the yield curve is more
highly inclined than the supply curve, the quantity of
water contributed from the source is in excess of the consumption; and, if there be storage-room available, this
excess may be impounded. Where, as at $a_2 b_1$, $b_2 c_1$, the
inclination of the yield-curve is less than that of the supply-
curve, the rate of supply exceeds the rate of yield; and
the storage-reservoir must be losing water, in order to meet
the draught on the mains for supply purposes.

If, now, lines $a_2 m$, $b_2 n$, $c_2 q$, parallel to the supply-curve,
be drawn touching the yield-curve at the culminating points
(points of inflexion) a_2, b_2, c_2, the total amount of such loss
from the reservoir is at any time measured by the difference
between the ordinate of the curve OA and that of the

N

parallel curves $a_2 m$, $b_2 n$, $c_2 q$, at the same point; and the loss is greatest where this difference is found to be greatest.

This maximum difference, which is the object of the investigation, can be readily found by simple inspection of the diagram, and is $c_1 n$, at the point c_1, in the example presented in *Fig. 29.* The amount of this difference, in whatever units it may have been expressed for convenience of plotting, is the absolute storage-capacity required to balance the yield and supply, in any period similar to the one investigated.

In this example, the quantity is 2×10^6 cubic feet; whence, the total supply being 4,500,000 cubic feet per annum, the required storage-capacity is

$$\frac{\dfrac{2 \times 10^6}{1}}{\dfrac{4\cdot5 \times 10^6}{365}} = 162 \text{ average days' supply.}$$

191. The diagram, *Fig. 29*, shows also the storage-capacity required, to utilise to the utmost extent the yield during the period considered, at a constant rate of supply— an important matter in schemes for the utilisation of water-power.

Suppose the rate of supply to be such that the greatest depression $n c_1$ of the curve $O A$ below the tangential lines, becomes exactly equal to its greatest altitude $c_2 d$ above the supply-curve. Then the latter attains a higher inclination, as indicated by the short dotted length $c_1 d$, and either of these equal quantities measures the necessary storage-capacity to utilise all the available water in the epoch considered. For $n c_1$, being equal to $c_2 d$, the reservoir would be, although full at c_2, emptied at c_1; and no greater rate of supply could be maintained, notwithstanding that the aggregate yield $A X$ during the period under review exceeds the aggregate supply $B X$. Hence the quantity measured by $n c_1$ or $c_2 d$ is the storage-capacity required to utilise the yield to the fullest extent, at the constant rate of supply indicated by the dotted supply-curve $c_1 d$ in the figure. This example may be sufficient to show that, although, doubtless, over an indefinitely long period of time, the yield and supply from any source might be equalised by sufficient storage ; such is by no means necessarily the case if the period under review be short, i.e. provided that a uniform supply is contemplated.

192. The process described is applicable to the investigation of every case of balance between an ascertained rate of entry of water into, and issue out of, tanks of all kinds, and might be employed to find the proper dimensions for service-reservoirs generally. The practical consideration, however, that such reservoirs exist not only to regulate periodical variations of consumption, but, further, to maintain a reserve of water to meet accidental stoppages of the supply from the source—whether these arise from the fracture of mains, the break-down of machinery, or the need for special cleansing—and that the provision which it is prudent to make against such accidents is usually large in comparison with the fluctuations of consumption, leads us to appeal directly to past experience of waterworks maintenance in determining the capacity proper to these structures.

The introduction of service-reservoirs into systems of water-supply arises from considerations of economy in the first place, and security in the second. In order to regulate and equalise the hourly and daily variations of water-consumption, and to provide against the contingencies of accidental stoppage of the supply, or its abnormal temporary increase due to conflagrations, all works for the

N 2

supply of water must be large enough to ensure the maintenance of the maximum supply, and should be duplicated to provide against accident, or disuse of the works during repairs and cleansing. It is, however, found generally the cheaper course to avoid the construction of the very large pumping-plant and supply-mains, duplicated for safety, which would be necessary to meet the maximum demands of consumers during the busy hours of the day or week, by the provision, near the centres of distribution, of service-reservoirs, into which the average supply from the source may be constantly delivered, and from which the fluctuating demands of consumption and occasional abnormal draught may be satisfied.

For a gravitation supply by a single main, three to six days' storage-capacity is desirable in the service-reservoir—depending upon the length, accessibility, and size of the supply-main. Where the length is short, it may prove more economical to duplicate the main than to build a reservoir. For a pumping-supply, one or two days' storage is generally ample ; and as the pumping-machinery is usually in duplicate, less than this amount of storage is not infrequently provided. Although, as a safeguard against fires, the adoption of the larger quantity is to be recommended.

The sizes of cisterns for house-supply, where it is necessary to use them, may be arrived at on precisely similar lines. To make them too small leads to inconvenience, if not, at times, to absolute famine ; whilst, if they are too large, the households suffer the risks of the pollution and natural deterioration of water, to which storage in small quantities upon inhabited premises renders it peculiarly liable.

193. The capacity of a reservoir having been determined, the next question in its design turns upon its site and the

materials to be employed in constructing it. In the selection of sites, the engineer naturally chooses positions which, whilst sufficiently elevated to ensure the gravitation of the water to the district to be supplied, and a proper pressure throughout the entire system of distributing-pipes, appear to lend themselves to the impounding of the required volume of water with the greatest facility and economy of construction. The advantages gained in these respects by raising the level of an existing lake, as, for example, in the works of the Manchester Corporation at Thirlmere, now in course of execution, is obvious ; and the formation of Lake Vyrnwy, in North Wales, by the construction of a dam across the narrow neck of the ancient glacial outlet from the valley, affords another example of the art of availing oneself of the advantages presented by nature in this respect.

The requisite elevation of a reservoir depends upon the pressure, or head of water, required by consumers at the place of supply, together with the portion lost in conveying the water from the reservoir to that point. These important considerations are fully discussed in the chapters on "distribution" and "conveyance" of water.

It need scarcely be observed that all proposed sites of reservoirs must be, as a preliminary step, carefully examined for the presence of any easily permeable rocks within their boundaries, as well as for faults and other dislocations of the strata, which might lead to either loss of the water or failure of any part of the structures erected to impound it.

In choosing positions for embankments, dams, or other heavy works, the foundations may be first tested by probings and borings ; but, before the works are designed, or can be correctly estimated as to cost, the ground must be thoroughly proved by sinking trial-pits into it, of size sufficient to admit of close examination of the strata upon which building is contemplated.

A compact rock or clay foundation may be built upon without special precautions; but fissured or permeable beds, gravel or alluvium, can only be rendered fit to serve as reservoir-floors or as foundations for the heavy works employed to impound water, if they are covered by a sufficient thickness of impervious material. In cases where permeable beds have an outcrop in the otherwise impervious basin of a river-valley, the site may be rendered available for a reservoir by carrying a puddled-clay or concrete wall across the beds in the line of the reservoir-dam—thus preventing any discharge of water under that structure. In clay foundations, it is desirable to prove the continuity of the clay strata under dams and embankments, by boring down a sufficient depth into it at a number of points—such boreholes being afterwards carefully plugged.

194. The next question of design—that of the materials to be used in the construction of reservoirs, is so intimately connected with their position, that it not infrequently exercises a certain amount of influence in determining the latter. If clay and earth are available in sufficient quantity in the immediate vicinity of the works, it may prove to be an economical course to impound the water by means of an earthen embankment. The absence of earth and an abundance of rock in the neighbourhood, among other considerations, sometimes suggest a masonry or concrete dam as the most economical—it necessarily is a highly durable—kind of structure. Brickwork is often well employed in the building of service-reservoirs; whilst iron and steel are generally the most suitable materials to use for small-sized and elevated tanks, and slate or iron for house-cisterns.

195. Having settled the position for a reservoir, the entire site should be closely contoured at every foot of

elevation, in order to determine the extent of the works required to impound the desired quantity ˙of water, the variation of water-level corresponding with different volumes in stock, and the area of the water-surface at every contour, when a natural depression in the ground is to be utilised for this purpose.

EARTHEN EMBANKMENTS.

196. We proceed to describe in outline the several classes of works employed in the storage of water. An earthen embankment generally consists of a mound of soil, of trape-zoidal section, supporting along its central line a vertical wall of impervious material, such as puddled clay or concrete. Sometimes the nature of the material available for the embankment is itself so fine and retentive, that the central wall is rendered unnecessary; but, where it is used, the integrity of the entire structure depends primarily upon the impermeability of this wall—the duty of the banks on each side being to afford lateral support to it, and to resist the thrust of the impounded water.

Well-puddled or tempered clay, if prevented from losing the water with which it is combined, forms a peculiarly tough, elastic and impervious body, capable of resisting the percolation of water under considerable pressures. Clay is a hydrated silicate of alumina, containing a variable amount of gelatinous or plastic silica. To be suitable for use as " puddle," it should be free from sand and vegetable matter, and should not contain soft friable stone ; but the presence of a certain amount of gravel in it is advantageous, and may be necessary.

The proper thickness for a puddle-wall depends upon the head of water to be retained by it, and the quality of the puddle. In practice, there is as much variation in the dimensions given to such walls, as there is variety in the constitution of different clays. For general guidance, it may be regarded a safe rule, with good puddle, to make the thickness of the puddle-wall at the base of the embankment equal to one-third of the depth of the impounded water in the reservoir—battering both sides to such an angle as will give a minimum thickness of 6 feet of puddle at the top of the wall. It is unnecessary to carry the puddle-wall above the highest wave-level of the reservoir (§ 237), and it is undesirable for it to have anywhere a less covering than 3 feet to 4 feet of ordinary earth, to protect it from the evaporative influences of sun and wind.

The puddle-wall must be carried down through the subsoil, to form a water-tight joint with the impervious base of the reservoir. It is laid in a trench excavated in the ground, the sides of which, frequently battered equally but always reversely to the sides of the puddle-wall in the embankment, afford the requisite lateral support to the material.

Owing to the comparatively unyielding character of the sides of this trench, and to the superior compression and consolidation of the puddle as greater depths are attained, as well as to the increased resistance opposed by the adjacent soil to the passage of water—the reversed batter of the puddle-wall below the ground-surface is found to be no source of weakness, whilst it is, practically, most convenient. The exact batter required in any particular case, is, of course, dependent upon the depth of the trench and the permeability of the adjacent earth; and, in practice, the thickness of the wall at the base is seldom found to be less than one-half the thickness possessed by it at the ground-level.

197. Cases frequently occur that demand very different treatment from the foregoing, in respect of the impervious strata of embankments. The occasional construction of

the latter entirely of retentive earth, without any special core at all, has been already alluded to. Sometimes the impervious core has been formed of an intimate mixture of gravel, sand and clay;* occasionally it has been successfully made of peat ;† examples are not uncommon in which concrete or rubble-masonry has been used for the purpose, whilst, often, a core is dispensed with, and the impervious stratum is laid as an apron under the pitching of the inner slope of the embankment. Whatever special adaptation of form is used in any such impervious sheets of clay or similar material introduced into embankments, it is all-important to observe the necessity for preserving them in a moist condition—especially in the case of clay-puddle, which, once cracked, will never heal naturally. For this reason, the interior portion of an embankment is formed of a finer material than that constituting the general body of the work, specially selected and built up with a view to retaining in its capillary pores the moisture it derives from the atmosphere.

198. The dimensions of the earthwork on each side of the puddle-wall, or other impervious core, in an embankment, is regulated, in the first place, by the angle at which the earth in question will stand, i. e. by its angle of repose. With ordinary materials it is a common practice to make the slope on the inner or water side, which is saturated with water, 3 to 1 ; and that on the outer side 2 to 1. But the width of every high embankment is dependent upon another condition, which is that of the maximum stress to which its foundation may be prudently subjected. The batter of high embankments is therefore flattened as greater depths below the top are attained, benchings being often formed

* Fanning, 'Hydraulic and Water-supply Engineering,' p. 348—New York, 1884.
† Trans. Liverpool Eng. Soc., vol. xi. p. 32.

at each place where the slope changes, to prevent the violent scour that would otherwise take place on the surface of long unbroken slopes during heavy falls of rain.

The top-breadth of embankments is a very variable feature, lying generally between 15 feet and 30 feet, dependent to some extent upon the roadway accommodation desired along them. The height to which the work is carried above the highest water-level of the reservoir is determined by the exposure of its situation and by its size, and is seldom less than 10 feet in structures of any considerable magnitude. *Figs. 30* and *32* present typical embankment sections.

The frictional resistance to horizontal movement possessed by such embankments as are here indicated, is, except upon specially slippery foundations, sufficient to overcome the horizontal thrust of the water impounded by them ; whilst their rigidity in relation to the pressures transmitted through them is far too great to permit failure through distortion of the work. We do not here allude to the great vertical distortion, or settlement, that takes place both during and after the construction of embankments ; the process of which is supposed to have ceased, or become negligible, before they are called upon to sustain their full charge.

199. To every reservoir belong two indispensable accessories—the outlet and the overflow, or " waste-weir," upon the design and construction of which depend the utility and the security of the entire works. In the formation of embankments for impounding-reservoirs, the first step is to provide for the passage of the water discharged from the catchment-area across the line of the works during their construction. This may be done either by building an outlet-culvert at a low level first of all, and passing the water through it during the erection of the embankment and auxiliary works around it ; or by forming a " by-wash "

past the site of the works at a high level, and diverting the natural flow of the stream that is to be dammed up into this channel at some convenient point higher up the valley —the level of the by-wash being generally higher than the top of the intended embankment.

The latter plan is usually adopted when the outlet from the reservoir is constructed under the embankment. The difficulties arising from the association of the rigid barrel of the outlet-culvert, of brick, masonry or iron, with the yielding inseparable from earthwork and puddle, are so considerable as to render the plan of carrying reservoir-outlets through such embankments extremely hazardous. Hence the practice of constructing such outlets in tunnels through solid ground, clear of the embankments, where the stresses due to the settlement of great masses of earth are absent, and where leakage or failure of the outlet-pipes or culverts will not threaten destruction to the whole works, has much to commend it; * although it must be admitted that the disturbance of the strata cut through in pursuance of this method is a disadvantage, in addition to which the cost involved by it is considerable.

Whether carried under the body of the embankment, or through the solid ground in its vicinity, the outlet-culvert is to be made of sufficient size to permit the free passage of the heaviest flood that is to be anticipated as likely to occur during the construction of the works. As an approximate guide, the following formula may be used, in this country, to determine the area of the discharge-culvert or tunnel.

Area of culvert in square feet

$$= \frac{R_m}{50} \times \sqrt{number\ of\ acres\ in\ the\ watershed,}$$

where R_m is the mean rainfall in inches.

* For a discussion of this question *vide* Minutes of Proceedings Inst. C.E., vol. lix. pp. 37 *et seq*.

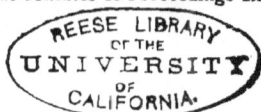

Tunnelled outlets are lined with the best brickwork or masonry in cement—scrupulous care being used to back the lining with good concrete so as to form a watertight joint all round between the shell of the tunnel and the natural ground. An outlet-culvert constructed across the site of the embankment itself is likewise built of brick or stone bedded upon concrete and covered with either concrete or puddled clay—preferably the former—throughout its length.

Its form is properly elliptical in cross-section, the shorter axis being horizontal and the eccentricity depending upon the nature of the soil in which it is built. Small culverts may, however, be made circular with advantage. In ordinary practical cases, ellipses whose axes have the ratio

$$\frac{\text{horizontal axis}}{\text{vertical axis}} = \frac{2}{3}$$

will satisfy the requirements of stability ; and, for convenience of access, the lower third of the section may be truncated and closed with a flat inverted arch, *Fig. 31.* The thickness of the wall of the culvert, if built of first-class materials, may be approximately estimated from the equation

$$t = b\sqrt{\frac{h}{2}},$$

where t is the thickness of the wall in inches,

   b is the smaller diameter of the culvert in feet,

and h is the height of the superincumbent earth in feet : the rule being understood to be merely applicable within the limits of ordinary practice.

200. The culvert, when constructed in a trench cut through the ground under the base of the embankment, thus obtains considerable lateral support from the natural ground ; but has a strong tendency, when loaded with the

enormous weight of the superincumbent embankment, to settle vertically in the middle of its length, especially at the place where it passes through the puddle-wall. Serious fractures may be, and unfortunately sometimes are, thus produced in it. Further, the material composing the puddle-wall has itself a tendency to settle, and so to part from the ~~lower~~ *under* side of the culvert at this place.

To meet these difficulties, two methods of construction have been employed :

(1) The culvert is constructed with a " slip-joint," i. e. a loose vertical joint is introduced over each side of the puddle-wall, so as to freely permit of unequal settlement there.

(2) A concrete or masonry pillar is built up from the bottom of the puddle-trench to support the culvert; in which, if the ground under the remainder of it is of a yielding nature, slip-joints should be introduced as in the first case. Neither method is free from objection ; although the latter is on many grounds the preferable plan.

201. When the embankment has been completed, and the outlet-culvert is no longer required to convey the natural discharge of the waters to be impounded past the site of works under construction, its permanent office is to contain the "draw-off-" or " supply "-pipe and the "discharge-" or " scour "-pipe, belonging to the reservoir.

These pipes are usually small in diameter compared with that of the outlet-culvert, which thus affords a useful means of access to them for examination and repair. Their actual dimensions are determined according to the considerations set forth in §§ 322, 323. They are sometimes passed through a strong brick or masonry " stopping," built across the culvert near to its inner end, and tied into it so as to form a water-tight diaphragm, that

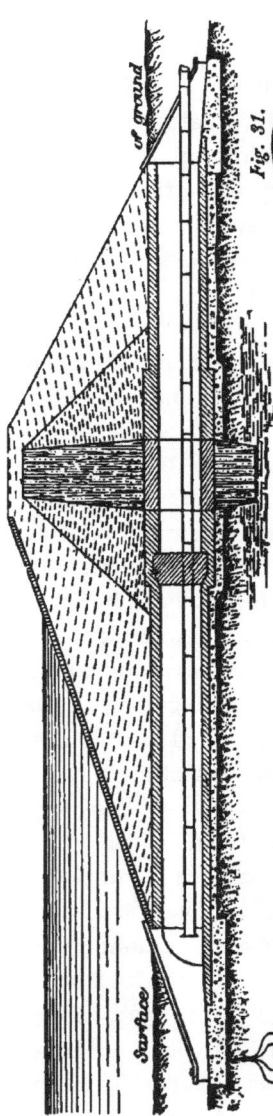

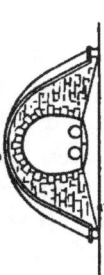

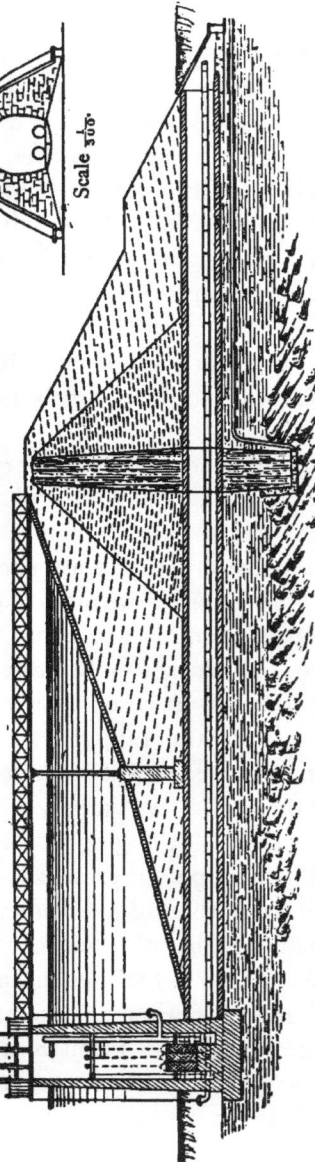

Fig. 30. Scale 1/100·

Fig. 31. Scale 1/100·

Fig. 32. Scale 1/100·

sustains the full pressure of the reservoir ; though, in the highest class of works, they are generally carried into a tower that is built over the inner end of the culvert, connection being made between them and the reservoir at several points in the tower (§ 245).

The former construction is illustrated in *Fig. 30*, which gives a typical cross-section of an embankment along the line of the outlet-culvert. The valves that command the "supply-" and "scour"-pipes are situated either in a chamber at the outer end of the culvert as shown, or at the bottom of a well sunk through the embankment in front of the puddle-wall. In both situations they threaten the security of the embankment, by permitting the existence of a charge of water at high pressure in the pipes under the whole or a portion of its width. A grid or screen over the inner end of the culvert, prevents the entrance of matter that would injure the valves or obstruct the pipes.

202. Both the inner and outer ends of the culvert are terminated by horizontal arches buttressed by strong wing-walls, forming respectively a " fore-bay " and "tail-bay," to resist the thrust of the embankment at these points during its settlement, *Fig. 31*. The strength required in these arches and wing-walls is considerable, and the stability of their foundations is of the greatest importance. If they are insufficient to sustain the thrust, the culvert is liable to be severely strained, and may be fractured by the spreading of the embankment during its settlement. The foundations must therefore be carried down to an unyielding base, and there built into it so substantially as to preclude all risk of slipping. The invert of the culvert may be carried along between the bases of the wing-walls, so as to distribute their thrust and at the same time prevent them from closing together. The wing-walls, if not spread too wide but carried to the toe of the embankment, approximately in the line of the

culvert, may be of moderate section. Each of them presents a case of a retaining-wall resisting the thrust of earth, the upper surface of which is practically level with the top of the wall. The more severely stressed of the walls are those of the fore-bay, which must sustain wet earth. The cross-section of the arches of both the fore-bay and the tail-bay may be made equal to that of wing-walls of corresponding height ; and the entire structure will then be possessed of ample longitudinal strength in addition to the transverse stability which is actually made the subject of investigation.

203. Retaining-walls are of such importance in water-works, being frequently necessary auxiliaries to the structures employed in them, that it may be advantageous here to briefly consider the case presented in § 202, remarking that, with some simple adaptations, the same treatment is applicable generally.

In presenting this investigation, we do not propose to assert that it is unexceptionable, or that the last word has been said upon the subject ; but a consideration of the purely mechanical aspect of the question, imperfect though our knowledge may be of the conditions of adhesion and friction obtaining in masses of earth, cannot fail to be an aid to intelligent design.

The condition of stability of a retaining-wall the base of which is prevented from sliding, is that the moment of the overturning couple formed by the thrust of the earth behind it into the distance of the centre of pressure from the base of the wall, shall be at least equal and opposite to the moments of the equilibrating couples of the wall itself. These couples are two in number, and comprise : (1) the weight of the wall into the distance of its resultant line of action from the toe, or outer extremity of the base of the wall ; and (2) the frictional resistance to vertical motion developed by the earth at the back of the wall into the breadth of the base.*

A primary assumption is, of course, rigidity of the several masses considered, and, consequently, a tendency of the wall to rotate about its toe. In determining the value of the earth-thrust against the back of the wall, an inherent difficulty lies in our ignorance of the coefficients of static friction, i. e. the tangents of the angles of repose in the interior of masses of earth. We therefore resort to the customary assumption that the coefficient of static friction in the interior of a mass of earth is equal to that of similar loose material at the surface—from which it follows that the resultant thrust of the earth against the wall acts at a point situated at one-third of its height above the base.

Fig. 33.

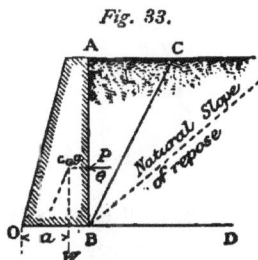

Considering a vertical section or slice of the wall with the earth behind it, one foot in thickness,— let h denote the height A B of the wall, b the breadth of its base O B, ρ the density of the earth, and μ its coefficient of friction (*Fig. 33*).

The thrust of the earth against the back of the wall may be regarded as due to the pressure of a prism of earth of triangular section separated along the line B C, making an angle θ with the vertical B A, from the whole adjacent mass.

The weight of this prism is $62 \cdot 4 \, \rho \, \dfrac{h^2}{2} \, tan \, \theta$; call this M.

Then the pressure normal to B C is $M \, sin \, \theta$, and the pressure parallel to B C is $M \, cos \, \theta$; therefore its resultant pressure in a direction parallel to B C is

$$M \, (cos \, \theta - \mu \, sin \, \theta).$$

* *Vide* Boussinesq, ' Essai théorique sur l'équilibre des massifs pulvéru-lents '—Paris, 1876.

O

Let P denote the force required to support the prism of earth if applied horizontally, i.e. parallel to B D ;

Then

$$M (\cos \theta - \mu \sin \theta) = P (\sin \theta + \mu \cos \theta).$$

Therefore

$$P = \frac{M (\cos \theta - \mu \sin \theta)}{\sin \theta + \mu \cos \theta} ;$$

or, substituting the actual value of M in this expression,

$$P = 62 \cdot 4 \, p \frac{h^2}{2} \times \frac{\sin \theta (1 - \mu \tan \theta)}{\sin \theta + \mu \cos \theta} . \qquad \text{(i.)}$$

From this it is evident that, for a certain value of θ, P has a maximum value.

Differentiating (i.) and equating to zero in the ordinary way, we obtain

$$\mu \, 2 \tan \theta = 1 - \tan^2 \theta ;$$

therefore

$$\mu = \cot 2 \theta,$$

or

$$\theta = \frac{1}{2} \cot^{-1} \mu .$$

Hence θ bisects the angle between the vertical B A and the slope of repose of the earth in question, whose coefficient of friction is μ ; and the value of θ thus found indicates the size of the prism of earth whose effective thrust against the wall is a maximum.

By assigning to θ the value $\frac{1}{2} \cot^{-1} \mu$ in equation (i.), the value of P is found, and it acts at $\frac{h}{3}$, above the point B, *Fig. 33.* By the assumption previously made, μ is the coefficient of friction between the earth and the wall ; let W denote the weight of the wall, and a the distance of the vertical through its *c. g.* from the toe O ; therefore, equating moments about O, the conditions of stability require that

$$P \frac{h}{3} = W a + P \mu b. \qquad \text{(ii.)}$$

From a solution of this equation, the proper dimensions of the wall are arrived at—the top-breadth being first chosen such as in practice is found to be consistent with durable construction in the situation considered. There are two other important factors of stability under the assumed circumstances : the first being that the resultant of P and W shall cut the base of the wall within the middle third of its breadth, which is equivalent to requiring that there shall be no tensile stress developed in it ; and the second being that every horizontal joint is possessed of sufficient shear-strength to resist the horizontal thrust above its plane. These matters are more fully considered in relation to masonry dams (§§ 221 *et seq.*); and it is sufficient to say that compliance with these two latter conditions is, in practice, generally involved in the observance of the rule expressed by equation (ii.).

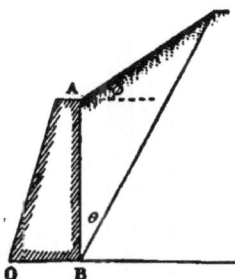

Fig. 34.

204. The investigation of the case of "surcharged" retaining-walls, which support a slope of earth, *Fig. 34*, may be followed out in a precisely analogous manner, substituting the value

$$62{\cdot}4\,\rho\,\frac{h^2}{2} \quad \cdot \quad \frac{\sin\theta\cos\phi}{\cos(\theta+\phi)}$$

for M in equation (i.)

205. When a tower is employed to contain the supply- and scour-pipes and the valves which form the working equipment of the outlet-works of a reservoir, it forms the inner abutment of the outlet-culvert. The variation of detail in the design of such works is dealt with in the numerous published descriptions of particular undertakings for water-supply. The typical example illustrated in

Fig. 35.

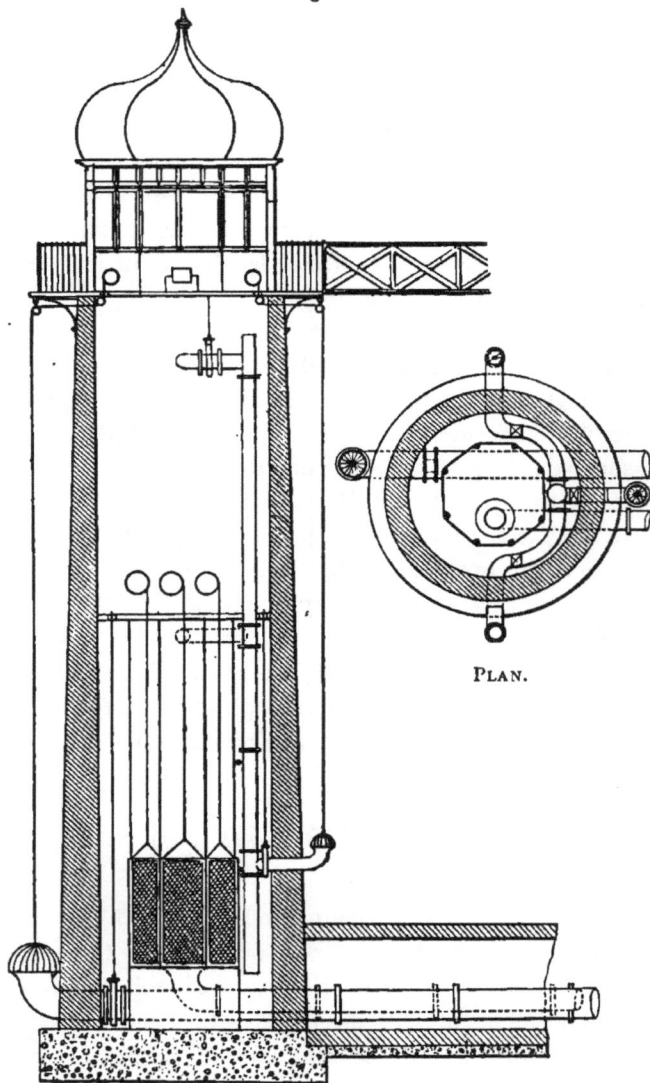

PLAN.

SECTION OF VALVE-TOWER.
Scale $\frac{1}{8}\frac{1}{1}$.

Fig. 35 is adapted to draw the supply from any one of three levels in the reservoir, as well as to scour or discharge the water from the bottom.

The supply-pipe is carried into the base of the tower within the screens; and draw-off branches, each commanded by a separate valve, communicate with a down-pipe outside the screens. The scour-pipe passes through the base of the tower. The straining-screens are carried in frames that can be raised at pleasure to be cleansed—duplicate screens taking their places during that operation.

The tower is founded upon a thick mass of concrete and is built of ashlar masonry or brickwork in cement. It is circular in plan, of a diameter dependent upon the space required for the straining-apparatus and the valves. The wall has a minimum thickness of 2 feet, and increases $\frac{1}{2}$ foot in that dimension for every 20 feet in depth below the top-water level of the reservoir.

206. In shallow and comparatively unimportant reservoirs, the supply- and the scour-pipes may, with proper precautions, be laid directly under the embankments. In such cases, the cost of a less exceptionable form of construction is sometimes prohibitive ; whilst the risk to the pipes from settlement or spreading of the embankment is inconsiderable—if built with due attention to its artificial consolidation.

The pipes are of specially strong section to support the superincumbent earth ; and are armed at the middle of the core of the embankment (whether of puddled clay or other impervious material) with circular shields termed puddle-plates, *Fig. 36*, made in segments and bolted together on the pipe at the required place, the joints being filled with lead and caulked. The consolidation of the core against such a shield is most effective in preventing " creep " of the water along the external surface of the pipe.

As in the case of culverts, pipes under such embankments must rest upon a strong homogeneous bed throughout their length.   If a suitable natural bed does not exist,
an artificial one may be formed of concrete, so as to prevent
the irregular settlement of the pipes
when subjected to the unequally-dis-
tributed weight of the embankment.
The pipes should run into a small
brick or masonry fore-bay and tail-
bay constructed on similar lines to
those described in the case of culverts
under high embankments.

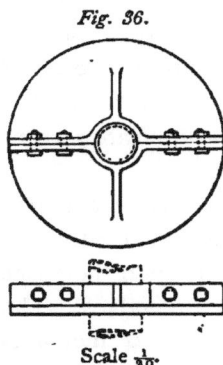

Fig. 36.

Scale $\frac{1}{30}$.

207. An alternative method of
drawing-off and discharging the water
from a shallow reservoir, avoiding the
risks attendant upon the existence of
passages through the body of the embankment, is afforded
by the adoption of siphon-pipes carried from the inner toe,
over the top of the puddle-wall, and terminating in a valve-
well at the toe of the outer slope, *Fig. 37.*   The siphons
are provided with valves at each end, the supply-siphon

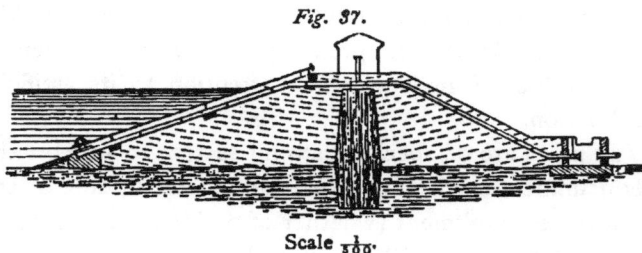

Fig. 37.

Scale $\frac{1}{300}$.

being further provided with one or two valves at different
levels ; the inner valves are worked by rods running in
guides down the slope of the embankment.

The pipes should run under the pitching of the inner,
and the turf of the outer, slope ; the several draw-off valves

and the valve-rod guides being set in blocks of concrete or in heavy stones.

The action of the siphon may be started by exhausting the air, and thus charging it with water by suction from its inlet-valve; or by closing all the valves and charging it directly through a cock at the top, then closing the charging-cock and opening the inlet- and outlet-valves—when flow will take place through the siphon in virtue of the difference of pressure at the inlet and outlet. The last-described is usually the more expeditious method of starting the action; but a small air-pump should be provided in connection with the charging-cock, for the purpose of exhausting the air that is disengaged from the water and collects at the top of the siphon, reducing its delivery by throttling the flow.

The tail-bay into which the siphons discharge may be formed as a circular well, with an overflow-weir of appropriate size, and an orifice for gauging the compensation water, if any. By a suitable arrangement of guides, the screens or strainers covering the valves may be drawn up to be cleansed and be restored to their positions, as required from time to time. It is of course essential to the complete action of the siphons that the discharge-well, or tail-bay, be so constructed that the level of the overflow-weir be below that of the lowest inlet-valves in the reservoir, by the amount of "head" necessary to produce the required velocity of flow through the siphons, and to overcome their frictional resistance (§ 318).

The questions of the size, strength and joints of pipes suitable for the outlet-works of reservoirs, as well as of the valves and other gear, are treated as forming part of the subject of the "Conveyance of Water," Chapter VI.

208. In preparing the foundation of a reservoir-embankment, the entire surface of the natural ground must be

divested of grass, roots and mould ; and that portion of it
which is to bear the impervious core of the embankment
must be excavated down to a continuous, firm and im-
permeable bed.

The soundness of this bed is a crucial feature of the
entire work ; and foundations are seldom found so naturally
perfect as not to require some artificial treatment in order
to render them satisfactory for their purpose. The passage
of water, in however small a stream, across any part of the
foundation of an embankment, will, if continued, by washing
away the friable material of which it is composed, inevitably
increase in quantity and ultimately threaten the ruin of the
entire work. Hence, in preparing the foundation of the
core or puddle-wall, pot-holes must be dug out and re-filled
with concrete, loose fragments of rock must be removed,
open joints must be cleared out and plugged with cement,
and springs must be led away in pipes and afforded free
egress on the lower side of the embankment, *Fig. 32.* All
this work must be performed with extreme care—with the
consciousness that, whatever tendency the foundation may
exhibit to yield water when laid open, will be enormously
accentuated under the pressure of the water impounded in
the reservoir when full, and that upward pressures are subtle
and dangerous in an exceptional degree.

All irregularities of the bottom should be removed.
Sudden variation of depth, and vertical faces in the foundation, are apt to produce rending of the structure during its
settlement—the tendency being for the whole mass to
coalesce towards its central base. Holes, joints and
fissures should be made good, as already explained ; rock
faces should be scarfed back ; and the whole foundation
prepared so as to be free from horizontal benchings and
steeply-inclined steps.

When a sound foundation is only to be met with at a

considerable depth below the surface of the ground, the work of sinking and of subsequently filling the puddle-trench, assumes an important aspect in regard to both the cost and the efficiency of the entire work. Hence, the prudent course is to spare neither time nor money in exploring, and ascertaining by actual trial, the fitness of a foundation, before going to the expense of opening it up completely.

209. In the case of the Woodhead reservoir of the Manchester Corporation, the actual excavation of the ground disclosed a foundation so different from what had been expected, that, although an attempt was made to use the site first selected, a new embankment had ultimately to be constructed in a different position.* At the Yarrow reservoir belonging to the Liverpool Corporation, similar conditions led to the sinking of the puddle-trench in " Turner's Embankment " at great expense, to the depth of 168 feet below the surface of the ground.† On the other hand, it is on record that a proposal to construct a great embankment on a site selected after considerable inquiry in the Bleasdale district of Lancashire, received a decisive check from the evidence of borings which proved that no sound bottom existed at 100 feet below the surface.‡

210. Borings afford valuable aid in indicating generally the continuity of ascertained strata ; but too much reliance may easily be placed upon their results. They cannot be said to be ever thoroughly trustworthy guides, and in wet ground are generally misleading. It would be an easy, although invidious, task to point to deplorable examples of misplaced confidence in the results of borings. Before a foundation is accepted, its fitness should be conclusively

---

* Bateman, 'History of the Manchester Waterworks,' p. 111.
† Beloe, 'The Liverpool Waterworks,' p. 15—London, 1875.
‡ Trans. Liverpool Eng. Soc., vol. i. p. 156.

proved by means of trial shafts of such size as to permit an examination of the strata traversed to be made by the engineer personally.

211. If the ground is firm enough, the puddle-trench may be excavated with its sides sloped to the batter of the puddle-wall. Where the ground is unsuitable for this method to be pursued, a timbered trench is sunk. The construction of timber-work for this purpose in a deep trench, requires much practical skill. Not only must the enormous thrust of the sides be resisted, but the timber must be inserted without disturbing the adjacent ground, and in such a manner as to be readily withdrawn as the raising of the puddle-wall proceeds. The best, although most costly, plan is to excavate each successive graft of the trench between rows of thin sheet-piles driven as the excavation proceeds; putting walings and stretchers in as each frame of timber is completed, in the customary manner—in-setting the successive rows of sheeting a little, to accord with the proposed batter of the puddle-wall.

212. The clay from which "puddle" is to be made should, during its excavation, be carefully freed from the presence of roots and all other vegetable matter; and should be spread out and exposed to the weather for as long a period as possible, before being used. The action of rain, frost and sun is to completely disintegrate the masses of clay—a treatment analogous to that of laying land fallow in agricultural work. This operation is termed "souring" the clay.

In constructing the puddle-wall, the foundation having been first carefully dressed and prepared, the "soured" clay is laid in a horizontal layer about nine inches deep on the lowest part of the bottom of the trench. A transverse grip is cut into it, into which water is poured. A gang of labourers working in line cut the clay with "puddle-spades" (narrow flat spades with blades 15 inches long), and draw it towards them, band by band, across the grip—thus forcing the water through it and treading it at the same time. The grip must from time to time be replenished with water, the amount required varying with the condition of the clay and the state of the atmosphere. This operation is termed "cutting" the clay, and is repeated two or three times in directions at right-angles, without the further addition of water after the desired plastic condition of the mass is attained. The clay is added layer by layer and is treated in a similar manner, each successive layer being incorporated thoroughly with that preceding it—the object of the long blades of the puddle-spades being to cut through the top layer of clay into that beneath it; and so, by sufficiently working it, to unite the whole into one homogeneous mass. In timbered trenches, the timber is withdrawn as the puddle-wall is raised, care being used to pack with clay any voids disclosed behind it.

213. When the puddle-wall has been raised to the ground-level, the construction of the earthen embankment is begun. The inner portion, made of fine, selected material, should be built in layers sloping inwards, punned in thin layers, or consolidated by traffic over it in layers not exceeding two feet in thickness—the puddle-wall being brought up slightly above it, and the greatest care being used that, in so doing, earth may not be carried from the embankment on to the puddle. The outer, coarser portions of the embankment are tipped in layers of three feet or four feet in thickness, following up the progress of the inner portion of the work, the slopes being kept full in order to allow for the effect of settlement. The embankment should be raised as far as practicable simultaneously from end to end, so that no portion of the surface may

exceed nine feet in height above the surface of any other portion in course of construction.

It may be observed that clay or fine earth, although, when moist and consolidated, well-adapted to resist the percolation of water, is ill suited for the purpose of forming an entire embankment, on account of its small angle of repose when saturated with water ; hence the necessity for the coarser material employed in the construction of the exterior portions of embankments.

214. When the work has completely settled—after a lapse of time that depends upon the rate at which the embankment has been raised, and the amount of artificial consolidation performed upon it during its construction— the inner slope is covered with a facing of stone, 15 to 18 inches deep, hand-pitched upon a layer of small stone scabblings or gravel. The object of this is to preserve the slope from the ravages of vermin and from the destructive effects of the wash of the waves, which, in some exposed reservoirs, attain considerable magnitude. The outer slope is sodded, or sown with grass seeds. Where it is necessary to make a road along the top of the embankment, the top-breadth should be considerable, the boundary-walls being well away from the crests of the slopes. The road should be kept in first-rate repair, in order to prevent surface-water from finding its way through the embankment to any considerable extent ; and provision should be made to prevent the passage of any extraordinarily heavy traffic along it.

215. The height to which it is practicable and desirable to build embankments of earth is dependent upon many considerations of the quality and quantity of materials available for the purpose, and of the character of the foundation. One of the finest examples of such structures in this country forms the Upper Barden reservoir of the Bradford Waterworks, constructed by Mr. A. R. Binnie. This

embankment has a height of 125 feet above the bottom of the valley.

216. The second indispensable accessory of a reservoir— the "waste-weir"—has for its object the prevention of a rise of the surface-level of the impounded water above the top of the impervious portion of the embankment.

By discharging the surplus water due to floods or other sudden increment of the contents of the reservoir, and so preventing the over-topping of the embankment, and its consequent immediate destruction by the scour of the escaping water, the "waste-weir" performs the office of a safety-valve, and, as such, demands the greatest attention in regard to its design, construction, and subsequent main-tenance. It may be safely said that many of the great reservoir disasters on record have been due to inadequate waste-weir accommodation.

The dimensions of the waste-weir are determined by two factors: (1) the maximum flood discharge; and (2) the depth of the stream of water which it is permissible, with due regard to economy in construction, to pass over the crest of the weir. This depth is generally limited to two feet as a maximum; and, with a free discharge from the crest of the weir, the length for the latter may be ascer-tained from the formula

$$l = \frac{Q}{8};$$

where l denotes the length of the waste-weir in feet, and Q denotes the maximum flood discharge of the catchment-area in cubic feet per second.

If, as frequently happens, proper flood-gaugings of the district under consideration are not available, the length of the waste-weir may be approximately determined from the equation

$$l = 2 R_m A^{\frac{1}{2}},$$

where R_m denotes the mean rainfall in inches, and A denotes the number of thousands of acres in the catchment-area. Where the proportion of reservoir-capacity to watershed-area is considerable, a smaller weir may suffice, owing to the balancing effect of the reservoir in retaining floods and thus mitigating their intensity.

217. The waste-weir is constructed, where possible, on the natural ground, of heavy masonry set in cement—the sill being composed of stones often as much as three feet in breadth and four feet in depth, and being provided with a concrete apron on its inner face. From the outer face of the sill, a flight of stairs between side-walls conveys the overflow past the embankment down to the stream or ditch that forms the natural artery for the drainage of the district. This structure, termed the "waste-water-course," must be strongly built upon a concrete foundation, if the natural rock cannot be conveniently reached by it. The stairs are formed of heavy blocks of stone, the ratio of rise to tread being such as to prevent any acceleration of the flow beyond that attained by it in falling over the sill of the waste-weir. The rise of each stair should not exceed two feet; but the design generally is largely dependent upon the nature of the foundation and the materials of construction available.

218. Owing to the strain to which the whole structure is subject during floods, and the frequent impossibility of effecting any repairs to it for long periods of time, the materials and the workmanship alike must be of the most substantial character. No temporary or movable parts can be safely admitted here. As an illustration of the kind of accident that may be confidently expected if these precautions are neglected, we may allude to an old experience at the Glencorse reservoir of the Edinburgh Waterworks. About the year 1850, permission had been obtained to raise the waste-weir of this reservoir one foot, so as to

increase the storage-capacity. This was effected by the addition of a strong balk of timber to the crest of the weir; and when, twenty years later, the removal of the balk was under consideration, it proved on thorough inspection to be so unsound that its removal was effected without loss of time—to prevent the catastrophe that would assuredly have taken place had the balk given way during a heavy flood.*

219. The most effective form of weir for any given length and section of crest is arrived at by placing it simply at right-angles to the direction of the current approaching it. Curved or slanting weirs, placed across a channel with a view to provide greater length and consequently less depth of over-fall, produce concentration of the current towards those parts of the weirs which are furthest up-stream—with diminution of the flow at the more remote down-stream parts. This is a result that might be expected to follow from the fact that the surface-slope towards any part of the weir is governed by the distance of its crest from the nearest cross-section of the channel where the flow is steady and the surface of the water is level.

220. When the impounding of water by embankments or dams affects streams that have considerable value as fisheries and spawning-grounds, it is necessary to provide for the passage of the fish into the reservoirs. The old form of salmon-stair, or ladder, consisting of a series of little pools formed as a flight of steps down which water flows in a cascade, is sufficient to meet cases of obstruction formed by low weirs; but, applied to high embankments, the necessarily flat inclination of such stairs would often render their extent inconveniently great. To meet this objection, fish-passes, on the lines of the well-known Mac-donald fish-way, are found serviceable. Essentially, the apparatus consists of a broad shallow trough, in which a

* *Vide* 'Report of the Parliamentary Enquiry into the St. Mary's Loch Water Scheme for Edinburgh, 1871.'

diaphragm parallel with its bottom runs from end to end at mid-depth ; the lower half thus forming a box, which is divided by a series of parallel diagonally-placed partitions, the spaces between them being connected with the upper trough at the ends. A series of half-spiral tubes is thus formed, and the water flowing down them and across the trough, even if the latter is inclined at a high angle, moves slowly in virtue of the small inclination of the tubes, up which the fish can readily pass.

MASONRY DAMS.

221. For the sake of presenting a connected account of the class of works necessary for an impounding-reservoir, we have, in the preceding sections, dealt somewhat fully with earthen embankments and the accessories belonging thereto ; but, as has already been pointed out (§ 194), such structures represent only one type of impounding-works.

The design of a masonry or concrete dam is not, like that of an earthen embankment, dependent largely upon the viscosity of the material of which it is constructed. The cohesion of the whole body is, if it be well built, so great that, assuming the pressures occurring in and transmitted through the mass of the structure to be everywhere less than some pre-arranged quantity, the question of the proper shape to be assigned to it might, if the law of its elasticity were completely known, be strictly investigated as a matter of rigid dynamics. The important questions relating to the foundations of dams and the materials and workmanship employed in their construction, affect the data upon which the design is based, and may be considered separately.

The form to be aimed at is one in which economy of material and workmanship is combined with such disposition of the parts of the structure that every one of them

may be applied to good purpose, in resisting either percolation or the thrust of the impounded water. Generally, it
is necessary for the foundation upon which the dam is
erected to be strong enough to bear the load imposed upon
it without yielding, and it is desirable that it should be
impervious to water. A rock foundation, firm and durable,
is pre-eminently suitable, as less likely to lead to unequal
settlement of the structure than a foundation of clay,
although the latter might be equally impervious, and, when
consolidated, just as well able to bear the load. Fuller
reference is made to this subject in §§ 208 and 256.

222. The complete solution of the question of the
stability of dams would require a knowledge of molecular
statics which has not, up to the present time, been forthcoming. But, as we shall show, under certain conditions, a
form of dam may be designed at once strong, stable, and
economical as to material.

The hypotheses under which we proceed to seek this
solution of the question are : (1) that the dam is a homogeneous rigid body,* and (2) that the stresses transmitted
through such a rigid body are either uniform in intensity,
or uniformly varying from point to point. Under the
latter hypothesis, the sum of the
stresses acting in one direction on
the line A B, *Fig. 38*, in a given
plane, if zero at A, and attaining
a maximum value, B C, at B, may
be represented graphically by the
triangle A B C. The resultant of
these stresses, which may be considered positive here, is
equal to their sum, and acts at the centre of gravity of

Fig. 38.

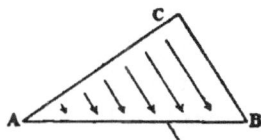

---

* The student of engineering will recognise here one of the many hypotheses that our ignorance of the properties of matter obliges us to adopt pro-
visionally as bases for investigations.

P

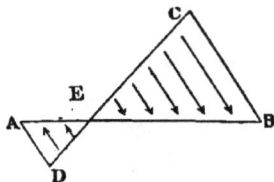

A B C in a direction parallel to any one of them, and it therefore cuts A B at a point distant one-third of its length from B. If there is a negative stress at A, whilst that at B remains positive, the resultant may be represented as the difference of two triangles, A D E, B C E, *Fig. 39*, and is positive or negative, and cuts A B at less than one-third of its length from B or A, according as the stress at B or at A is numerically the greater. If the stresses at both A and B are positive, the resultant cuts A B at more than one-third of its length from B or from A, nearer to the end at which the stress is the greater, and, in the case in which those stresses are equal, bisects A B.

Fig. 39.

From this it follows that, whether the maximum positive stress occurs at A or at B, the essential condition of the absence of negative stress in A B is that the resultant of all the parallel stresses acting on it shall be numerically equal to their sum, and shall not cut it outside of the middle third of its length.

223. Structures of concrete and masonry offer such high resistance to compressive stress, and are so variable and uncertain in respect of tensile strength, that it is the usual practice to rely only upon their resistance to compression in estimating their stability. Compressive and tensile stresses being regarded as positive and negative respectively, the results of the foregoing investigation may be directly applied to them. And here it may be worth while to advert briefly to the benefits to be derived from refined examination into the statics of such bodies as masonry dams.

In many classes of engineering structures, a study of existing types is of itself frequently sufficient to lead to the evolution of designs appropriate to given circumstances.

But in the case of dams, engineering practice exhibits a want of uniformity which must in many cases be ascribed to an inadequate conception of the principles that should govern their design, no less than to incomplete information as to some of the important physical properties of the materials of which they are built.

Without doubt, the vastness of the forces to be contended against, and the fearful risks attending rupture or other failure of dams, contribute to produce a demand for structures whose mere mass may be viewed with complacency by the public. But, when those massive proportions are attained at the cost of bringing into action excessive stresses in the structure with an unnecessary and lavish expenditure of capital, there can be little question that the engineer should disregard the popular view and follow the guidance of scientific principles, the application of which in this branch of construction, was, to the credit of French engineering talent, first demonstrated at Furens and the Ban to lead to safe results in practice.

224. The essential conditions of the strength and stability of a dam are:

(1) The maximum (compressive) stress, in either the inner or the outer face of the structure, shall not exceed in intensity a safe limit—found in existing works to be between 6 tons and 12 tons per square foot.

(2) There shall be no tensile stress at any part of the structure. This involves (§ 222) that, under all conditions of loading, the resultant stress shall not in any plane vertical section fall without the middle third of the breadth of the dam.

(3) The resistance to shear in any horizontal plane shall be greater than the total horizontal thrust of the water at the level of the plane considered.

(4) The structure shall be impervious to water.

225. Observance of the first two conditions stated in § 224 generally involves the fulfilment of the third. The fourth condition is never absolutely complied with ; for even the most compact and homogeneous rock is permeable in some degree. Still, by skilful selection of the materials of construction, and sound workmanship, any sensible amount of flotation or "up-lift" of the structure may be prevented, by which the important element of its weight would be affected.

It is, further, of the utmost importance that the requisite strength and stability of a dam be obtained without excessive cost—which generally implies without excess of material. Hence, a fifth element of a perfect design is that the primary conditions of strength and stability shall be fulfilled with that disposition of material which gives the smallest cross-sectional area in any vertical plane.

226. The two classes of dams to which allusion has been already made (§ 119) are: Those which are straight in plan and depend for their stability upon their power to transmit the thrust of the impounded water to the foundation upon which they rest (sometimes called "gravity" dams) ; and those which, arched in plan, transmit the thrust of the water to their abutments on the flanks of the hill-sides between which they are usually situated. The latter class are generally known as "arched" dams.

227. The nature of the general problem presented in the preceding sections is highly complex, and no complete solution of it has yet been given, although the investigations of Graëff, Delocre, Rankine and others have resulted in more or less close approximation to the truth within certain limiting conditions.

There is, however, one case in which a rigid determination of the problem is readily obtained ; and, as a clear conception of the statical conditions of the entire question and the general mode of treatment follows from a consideration of this case, we propose, notwithstanding that it cannot be completely realised in practice, to examine it.

228. CASE I.—To find the most economical form of section for a dam required to sustain still water up to the level of its crest, and to conform to the primary conditions of strength and stability.

Let ρ denote the density of the masonry referred to that of water as unity.

Using Cartesian co-ordinates x and y in the ordinary way, the origin being taken at the crest of the dam, and considering a plane vertical section of unit thickness : (*a*) when the reservoir is empty, the resultant of the vertical forces arising from the weight of the structure, is, at any horizontal joint, $x_1\,y_1$, *Fig. 40,*

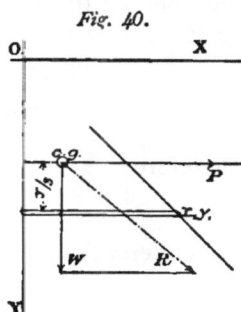

Fig. 40.

$$\rho \int_0^{y_1} x\,dy. = W.$$

Call this W.

In its limiting position W must cut the joint at one-third of its breadth, i. e. $\dfrac{x_1}{3}$, from the inner profile of the section.

(*b*) When the reservoir is full to the crest, the resultant of the weight of the dam and the thrust of the water above the same horizontal joint must, in its limiting position, cut the joint at two-thirds of its breadth from the inner profile.

The thrust of the water above the joint considered is

$$\int_0^{y_1} y\,dy,$$
$$= \frac{y_1^2}{2}. = P$$

Let this be denoted by P.

The line of action of P is horizontal, and is situated at a distance $\dfrac{y_1}{3}$ above the joint. By the triangle of forces (*Fig. 40*)

$$\frac{\rho \int_0^{y_1} x\, dy}{\dfrac{y_1^2}{2}} = \frac{\dfrac{y_1}{3}}{\dfrac{x_1}{3}} ; \quad \checkmark$$

therefore

$$x_1 \int_0^{y_1} x\, dy = \frac{y_1^3}{2\rho}. \qquad\qquad \text{(i.)} \checkmark$$

But, finding the centre of figure with reference to the axis OY,

$$\frac{x_1}{3} \int_0^{y_1} x\, dy = \int_0^{y_1} \frac{x^2\, dy}{2},$$

therefore

$$\frac{y_1^3}{2\rho} = 3 \int_0^{y_1} \frac{x^2\, dy}{2}, \qquad\qquad \text{from (i.).}$$

Differentiating,

$$\frac{3\, y_1^2}{2\rho} = \frac{3\, x^2}{2},$$

therefore

$$x = y \sqrt{\frac{1}{\rho}}. \qquad\qquad \text{(ii.)} \cdot$$

Hence, the required section is a triangle, the inner side being vertical, and the breadth at any given depth is determined by equation (ii.) as far down as the point where the limiting intensity of stress is attained in the outer face of the dam.

229. Such a triangular profile satisfies completely conditions 1 and 2, § 224. The thrust of the water being $\dfrac{y_1^2}{2}\, w$, and the weight of the dam being $\dfrac{y_1^2}{2}\, w \sqrt{\rho}$ above the

joint considered, w denoting the weight of a cubic foot of water; condition 3 requires that the coefficient of resistance at that joint shall be not less than $\sqrt{\dfrac{\bar{I}}{\rho}}$, which is generally, although it has not been invariably, the case in practice.

Proceeding further, it is desirable to ascertain the depth below the crest or top of the dam, to which the triangular profile may be extended without causing the occurrence of stresses exceeding some prescribed limit of intensity, which is known from experience to be permissible in the class of masonry dealt with.

Let s denote this limiting intensity of stress, and let a denote the vertical angle of the section,

$$\text{i.e.} \quad a = \tan^{-1}\sqrt{\frac{\bar{I}}{\rho}};$$

Let W denote the total weight of the dam on its base at any depth h below the crest; then, if the reservoir is empty,

$$W = \frac{h^2}{2} \cdot \sqrt{\frac{\bar{I}}{\rho}} \cdot \rho w;$$

and the limiting value of y is reached when

$$\frac{\dfrac{h^2}{2} \cdot \dfrac{\rho w}{\sqrt{\rho}}}{\dfrac{h}{\sqrt{\rho}}} = \frac{s}{2};$$

that is, when

$$h = \frac{s}{\rho w}.$$

230. If the reservoir is full, w⬛⬛⬛⬛e practical condition required, the limiting height⬛⬛⬛e triangular profile,

Fig. 41, being that at which the maximum stress-intensity
s is attained in a projection of the horizontal base on to a

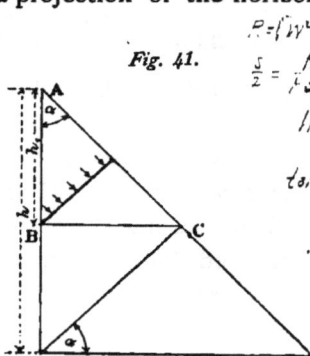

Fig. 41.

plane perpendicular to the resultant pressure, is $h \cos^2 a$;

therefore
$$h_1 = \frac{s}{\rho w} \cos^2 a$$

$$= \frac{s}{w (\rho + 1)}. \qquad\qquad \text{(ii.)}$$

EXAMPLE.—Take $s = 10$ tons per square foot;

$$\text{,,} \quad \rho = 2\cdot25,$$

and, consequently, $\quad a = tan^{-1} \dfrac{2}{3}$;

take $w = \dfrac{1}{36}$ ton;

then the foregoing triangular profile may be adopted for a
dam not exceeding in height

$$\frac{10}{\frac{1}{36} \times \frac{13}{4}} = 111 \text{ feet.}$$

Referring to equation (ii.) above, the inclined parallel
forces acting at any joint B C, *Fig. 42*, are considered to

act across a section normal to their direction, through which section they are naturally resisted. The *Fig.* indicates diagrammatically the action of the individual forces upon particles whose faces, normal to the direction of the forces, form in the aggregate the section of re-sistance. This section is equal to B C multiplied by the cosine of the inclina-tion of the forces to the vertical. Com-pare *Fig. 41.*

Fig. 42.

231. Dams which fall within the limit of height thus found, may be conveniently termed " low " dams ; whilst those in which the maximum stress-intensity occurs in the base, may be designated " high " dams. A special method of investigation must be applied to this latter class, the profiles of which are affected by the increase of width necessary to prevent the intensity of the stress at their bases from exceeding the given maximum value *s*.

232. The upper portion of the high dam, as far below the crest as $\dfrac{s}{w\,(\rho + 1)}$ (§ 230), assumes the form already investigated for the low dam ; but, below that depth, the stresses which occur, in relation to the resistance to crushing of the masonry, introduce a fresh element into the investi-gation of the profiles. The relation of the breadth of the dam to its height is now dependent not only on the position of the resultant stress, but also upon its amount ; and the determination of this latter quantity strictly involves a knowledge of the very relation that is sought. In fact, the general form and the mass of the dam are indispensable data for the calculation of its base. The problem thus calls for the application of some method of investigation suited to the treatment of indeterminate quantities.

Graphic methods may be employed, but, although very useful as a check upon the results, for fairly accurate work they are somewhat laborious, and involve the difficult and oft-repeated operation of finding the centre of gravity of a curvilinear area.

233. CASE II.—We are enabled to present a simple, practicable solution of the problem, complying with the four primary conditions of strength and stability, § 224, by introducing the following condition—

(5) The resultant of the weight of the dam at any horizontal base and of the water over its inner face, shall cut that base at a distance e from its inner face not less than one-third of the breadth of the dam in that plane.

The process of solution is based upon the following theorem :—

> If a rigid body rest upon a horizontal surface, the distance of the centre of pressure from the extremity of the base at which the maximum stress occurs is
>
> $$\frac{b}{3}\left(2 - \frac{s}{2\,s_1}\right),$$

where b is the length of the base upon which the body rests,
s is the maximum stress-intensity on the base,
and s_1 is the average stress-intensity on the base.

This theorem is easily proved, thus :—

Let y denote the stress occurring at any distance x from the extremity of the base : At this extremity the stress is s, and at the other extremity it is $2\,s_1 - s$; and

$$y = s - x\,\frac{2\,(s - s_1)}{b}.$$

Let c denote the required distance of the centre of pressure from the extremity of the base, then

$$c b s_1 = \int_0^b x y \, dx = \int_0^b s x \, dx - \int_0^b \frac{2(s - s_1)}{b} x^2 \, dx$$

$$= \frac{1}{2} s b^2 - \frac{2(s - s_1)}{b} \cdot \frac{b^3}{3},$$

$$\therefore \quad c = b \left(\frac{4 s_1 - s}{6 s_1} \right) = \frac{b}{3} \left(2 - \frac{s}{2 s_1} \right). \quad Q.E.D.$$

234. We now proceed with the investigation. Let W denote the total weight of the masonry and the water over the inner face of a vertical section of the dam of unit thickness; let P denote the corresponding horizontal thrust of the water against the inner face of the dam; and let R denote the resultant pressure on the base. Then $P = \frac{w}{2} h^2,$

and $R = \sqrt{W^2 + P^2}$ (*Fig. 43*).

Fig. 43.

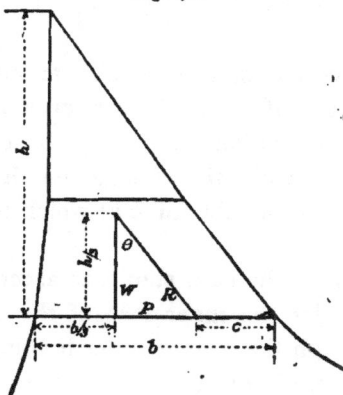

If s is the maximum intensity of stress in the dam, the maximum stress-intensity on a horizontal section is

$$s \cos \theta = s \frac{W}{\sqrt{W^2 + P^2}};$$

the average stress-intensity on b is

$$\frac{\sqrt{W^2 + P^2}}{b}.$$

Hence (theorem) $c = \dfrac{b}{3}\left(2 - \dfrac{s\,b\,W}{2\,(W^2 + P^2)}\right)$;

assuming e, condition 5, to be one-third of b,

$$b = \frac{b}{3} + \frac{P}{W} \cdot \frac{h}{3} + \frac{b}{3}\left(2 - \frac{s\,b\,W}{2\,(W^2 + P^2)}\right),$$

hence

$$\frac{P\,h}{W\,b} = \frac{s\,b\,W}{2\,(W^2 + P^2)},$$

$$\therefore \quad b^2 = \frac{2\,P\,h\,(W^2 + P^2)}{s\,W^2};$$

and, substituting the value $\dfrac{w}{2} h^2$ for P,

$$b = \sqrt{\frac{w\,h^3}{s}\left(1 + \frac{w^2\,h^4}{4\,W^2}\right)}. \qquad \text{(i.)}$$

From this equation, b is readily found with a high degree of accuracy, if W be known even approximately. A second approximation to b may be easily made if required, as W is the only variable on the right of the equation for any given value of h at which the breadth b is being computed.

Having found b, the next step is to ascertain how much of it must lie under the inner face of the dam, i.e. within the vertical through the crest, in order to bring the incidence of the resultant weight of the whole superincumbent mass of masonry and water to the distance $\dfrac{b}{3}$ from the inner toe.

This is arrived at by equating the moments of the vertical forces about the point in question, O_1 *Fig. 44*, to zero.

Thus, considering a lamina of unit thickness and of depth d, the upper and lower breadths of which are respectively b_0 and b_1, *Fig. 44*, situated at depths h_0 and h_1 below the crest of the dam, let W_0 denote the total super-incumbent weight on the upper surface of the lamina, acting at O_0, $\dfrac{b_0}{3}$ from the inner face ; and let x_1 denote the

Fig. 44.

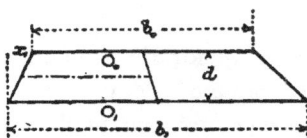

required increment of the breadth under the inner face of the dam, due to the addition of the lamina under conside-ration. Then the weight of the lamina is

$$\frac{b_0 + b_1}{2} \times \rho\, w\, d\,;$$

the distance of its c.g. from the vertical axis through O_1 is

$$\frac{3\,b_0 + 6\,x_1 - b_1}{12}\,, d \text{ being small};$$

the weight of water over the inner face of the lamina is

$$\frac{h_0 + h_1}{2} \times w\, x_1\,;$$

the distance of its c.g. from O_1 is

$$\frac{b_1}{3} - \frac{x_1}{2}\,;$$

and the moment of W_0 about O_1 is

$$W_0\left(\frac{b_1 - b_0}{3} - x_1\right).$$

The condition of equilibrium about O_1 of the entire mass above the base b_1 is expressed by

$$\left.\begin{array}{l} \dfrac{\rho\,w\,d}{24}\,(3\,b_0{}^2 - b_1{}^2 + 6\,x_1\,(b_0 + b_1) + 2\,b_0\,b_1) \\[2em] -\dfrac{w\,x_1}{12}\,(h_0 + h_1)\,(2\,b_1 - 3\,x_1) - W_0\!\left(\dfrac{b_1 - b_0}{3} - x_1\right) = 0, \end{array}\right\} \text{(ii.)}$$

which is an ordinary quadratic equation for finding x_1; whence the precise value of W_1 is fixed. The process of calculation may be repeated and applied to lower laminæ so as to determine values of b_2, b_3, and so on.

235. The method described in § 234 is rendered applicable to dams of any height, by making the relation of e to b, § 233, such that the value of s in the equation

$$s = \frac{2\,U}{b^2}\,(2\,b - 3f),$$

where U is the weight of the dam above the base b,

and f is the distance of its centre of gravity from the inner toe,

does not exceed the prescribed limiting stress-intensity.

236. In applying the principles of stability thus far developed to the design of dams in practice, there are two further circumstances that require consideration: (1) the top-breadth of the structure, and (2) the effect of wind-pressure on its outer face when the reservoir is empty.

The top-breadth of a dam is a feature that is governed by local conditions. The thin edge shown in *Fig. 43* is in practice inadmissible, because it could scarcely be built so as to be impervious to water, and it would not resist wave-action or the thrust of ice. Beyond this, a foot-path or carriage-way is almost always required along the top of a dam ; and the breadth may therefore be taken as varying necessarily between 5 feet and 10 feet; and, in some cases it is considerably more than this.

237. Having fixed the top-breadth, the next step is to discuss the effect of the addition of a triangular mass of masonry, such as A B C, *Fig. 45*, to the triangular section considered in § 228.

Fig. 45.

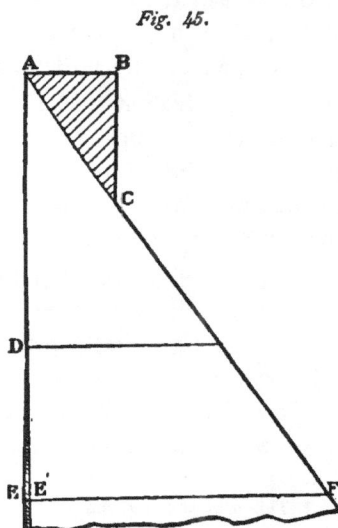

Still assuming the existence of hydrostatic pressure against the inner face up to the crest A—for the oscillatory waves produced in deep reservoirs may be assumed to exert approximately the hydrostatic pressure due to their height; and, generally, it is unnecessary to carry the body of the dam above the level of the crests of the largest waves in the reservoir, i. e.

$1 \cdot 5 \sqrt{D} + (2 \cdot 5 - \sqrt[4]{D})$ feet above top-water level; [*] where D is the "fetch," or longest line of exposure of water-surface to wind, expressed in statute miles—it may be easily shown that the conditions of stability permit both the inner and outer profiles of the section to be carried down vertically from the top to the point C, *Fig. 45*, which is situated at $b \sqrt{\rho}$ below B, where b denotes the top-breadth A B, and ρ is, as before, the specific gravity of the masonry. The special case of a dam acting as a weir and surcharged by water during floods is treated by measuring h (§ 229) from the flood water-level and not from

[*] Stevenson, 'The Design and Construction of Harbours,' 3rd edition, p. 29—Edinburgh, 1876.

the crest of the dam ; the point of application of P being of course slightly affected. The method of calculation is similar to that sketched out in the preceding sections.

238. The addition of the triangular mass of masonry A B C causes the centre of vertical pressures of the entire section (i. e. a vertical through its c.g.) to fall within the middle third of any horizontal joint down to a depth below the crest equivalent to $A D = 2 b \sqrt{\rho}$. Below the point D, the vertical through the c.g. of the section would fall in the inner third of every horizontal joint, that is, nearer to the inner face than one-third of the breadth, approaching that face down to E′F, which is situated at a depth $3 \cdot 1 b \sqrt{\rho}$ below the top of the dam.* Proceeding still lower, the incidence of the verticals on successive horizontal joints, would gradually recede from the face towards the middle third of the joints.

In order to avoid thus violating condition 2, a batter is given to the inner profile of the section, of amount

---

* This may be found thus :

Denote by $C : h \sqrt{\dfrac{I}{\rho}}$ the ratio that is borne by the distance from the inner face at which the vertical through the c.g. of any section cuts the corresponding joint or base, to the total breadth of that joint.

Required the value of h which makes this ratio a minimum.

By taking the moment of the area $A\,E'\,F\,C\,B$, *Fig. 45,* about a vertical axis through F, we have

$$C = \frac{2\,b^3\,\rho^{\frac{3}{2}} + h^3}{3\,(b^2\,\rho^{\frac{3}{2}} + h^2\,\rho^{\frac{1}{2}})} ;$$

and, by applying the ordinary process for finding maxima and minima values of this function of h, it will be found that the value

$$h = 3 \cdot 1\,b\sqrt{\rho}$$

gives a minimum value of the expression

$$\frac{C}{h\sqrt{\dfrac{I}{\rho}}} .$$

sufficient to maintain the line of resultant pressures,* reservoir empty, within the middle third of the section, down to a point situated $3 \cdot 1 \, b \sqrt{\rho}$ below the crest (E, *Fig. 45*). Below this point the inner profile may be vertical down to a depth

$$\frac{s}{w \, (\rho + 1)} \cdot \quad \text{(Equation (ii.) § 230.)}$$

To find the batter required,

Let $n = $ E E' (*Fig. 45*), the addition to the breadth of the dam under the batter D E. Equate the sum of the moments of the three triangular areas A B C, A E' F, D E' E, about F to the moment of their sum, acting at two-thirds of F E from F, about the same point. Hence is obtained for the value of n, for use in cases that arise in ordinary practice, approximately

$$n = \frac{b}{16} \cdot$$

The required batter is, therefore,

$$\frac{\frac{b}{16}}{1 \cdot 1 \, b \sqrt{\rho}} = \frac{1}{18 \sqrt{\rho}} \cdot$$

The effect of this modification of the elementary triangular form of section is to cause the line of resultant pressures, reservoir full, to fall slightly within the middle third of any given joint or base ; and the section is therefore not strictly of the most economical form possible. The defect is slight, and may if desired be corrected by a series of approximations ; but upon several considerations that will suggest themselves to practical men, the simple and highly economical form of section arrived at in the above investigation may be accepted without such correction—complying as it does with the primary conditions of strength and stability.

---

* By "line of resultant pressures" is meant the line joining the points of incidence of successive resultants upon corresponding horizontal joints of the dam.

Q

239. The action of wind-pressure upon the outer face of a dam introduces an element into the investigation of stability, which should, strictly speaking, be considered simultaneously with those of weight and water-pressure. The effect of the horizontal component of the wind-pressure is to throw the lines of resultant pressures nearer to the inner face of the dam, whilst that of its vertical component is to increase the intensity of the stresses upon the base. The total effect is so slight that it may be neglected in every case except that of the "high" dam, reservoir empty. Here it would, at a certain point of the inner face, if not provided against, violate the fundamental condition (1)— by causing the occurrence of stress in that face which would exceed the limiting value s. Such an effect is, however, provided against by condition 5, § 233, introduced to meet the case of high dams.

How slight a shift of the line of resultant pressures, reservoir empty, is caused by a considerable wind-pressure acting upon the outer face of a dam, may be seen from the following examination of the case, which will probably satisfy the reader that the course, frequently adopted in practice, of neglecting its consideration, is not unreasonable.

240. Considering the case of a dam of simple triangular section—the inner face being vertical and the outer (inclined) face being exposed to the action of wind-pressure; and, as the worst case, assuming the reservoir to be empty:

Let p denote the intensity of the wind-pressure against a vertical plane;

w the unit-weight of water;

ρ the *s.g.* of the masonry;

and a the vertical angle of the dam.

The wind-pressure normal to the outer face is $p \cos a$; the total horizontal thrust of the wind down to any depth h below the crest of the dam is $h\,p\,cos\,a$: the total weight of masonry in a section of unit-thickness, corresponding with this value of h, is

$$\frac{1}{2}\,h^2\,w\,\rho\,tan\,a.$$

The inclination of the resultant of the wind-pressure and the weight of the dam to the vertical is

$$tan^{-1}\,\frac{2\,h\,p\,cos\,a}{h^2\,w\,\rho\,tan\,a}\,;$$

which, substituting the value $\sqrt{\dfrac{1}{\rho}}$ for $tan\,a$ (§ 229), may be written

$$tan^{-1}\,\frac{2\,p\,cos\,a}{h\,w\,\sqrt{\rho}}.$$

From this it appears that the batter varies inversely as h. Therefore, as the point of application of the resultant wind-thrust above any given base, is situated $\dfrac{h}{2}$ above that base ; the amount by which the line of resultant pressures is forced out of the vertical by the thrust of the horizontal component of the wind-pressure, is constant at all depths, and is

$$\frac{p\,cos\,a}{w\,\sqrt{\rho}}.$$

The intensity of the vertical component of the wind-pressure is $p\,cos\,a\,sin\,a$, and its total amount is, at any depth h,

$$\frac{h\,p\,cos\,a\,sin\,a}{cos\,a}$$

$$=\,h\,p\,sin\,a.$$

Q 2

This force acts at the middle of the base, and causes a shift in the resultant vertical pressure nearer to the middle of the base by an amount

$$\frac{p \sin a}{3\,w\,\rho}.$$

Hence the line of resultant pressures is thrown nearer to the inner face of the dam by the amount

$$\frac{p}{w\sqrt{\rho}}\left(\cos a - \frac{\sin a}{3\sqrt{\rho}}\right).$$

EXAMPLE.—If $p = 56$ lbs. $= \dfrac{1}{40}$ ton per square foot,

$$w = \frac{1}{36} \quad \text{,,} \quad \text{,,}$$

$$\rho = 2\cdot25\,;$$

then

$$\cos a = \frac{5}{6},$$

and

$$\sin a = \frac{1}{2}\,;$$

therefore the shift of the line of resultant pressures, reservoir empty, due to the assumed wind-pressure on the outer face of the dam is

$$\frac{\dfrac{1}{40}}{\dfrac{1}{36}\times\dfrac{3}{2}}\left(\dfrac{5}{6} - \dfrac{\dfrac{1}{2}}{3\times\dfrac{3}{2}}\right)$$

$$= \frac{1}{5}\times\frac{39}{18},$$

i. e. $< \dfrac{1}{2}$ foot.

241. In illustration of the methods sketched above, we propose to add here an example containing a numerical solution of the several problems comprised in the process of design ; and to supplement it by an outline of the operation of applying graphic statics to the case in order to check the results obtained by calculation.

Consider a vertical plane section of the dam, 1 foot thick.

Let b, the top-breadth, $= 10$ feet,

ρ, the *s.g.* of the masonry, $= 2 \cdot 25$,

w, the weight of a cubic foot of water, $= \dfrac{1}{36}$ ton,

s, the limiting stress-intensity, $= 10$ tons per square foot :

Then the weight of 1 cubic foot of the masonry will be $\dfrac{1}{36} \times 2\dfrac{1}{4} = \dfrac{1}{16}$ ton ; and, since $\tan a = \sqrt{\dfrac{1}{\rho}}$, $\cos a = \dfrac{5}{6}$;

also $\sqrt{\rho} = \dfrac{3}{2}$.

Measuring depths parallel to the vertical axis A L through the crest, and breadths to the inner and outer profiles from that line, *Fig. 46,*

A B $= 10$ feet ;

A D $= 2 \times 10 \times \dfrac{3}{2} = 30$ feet (§ 238);

D E $= 1 \cdot 1 \times 10 \times \dfrac{3}{2} = 16 \cdot 5$ feet (§ 238);

A L, the limiting height for the low dam, is

$$\dfrac{10}{\dfrac{1}{36} \times \dfrac{13}{4}} = 111 \text{ feet (§ 230) ;}$$

L M $= \dfrac{111}{\dfrac{3}{2}} = 74$ feet (§ 228, ii.) ;

$$B\,C\ =\ 10\ \times\ \frac{3}{2}\ =\ 15\ \text{feet}\ (\S\ 237)\ ;$$

$$E\,E'\ =\ L\,N\ =\ \frac{10}{16}\ =\ 0\cdot 6\ \text{foot}\ (\S\ 238).$$

Fig. 46.

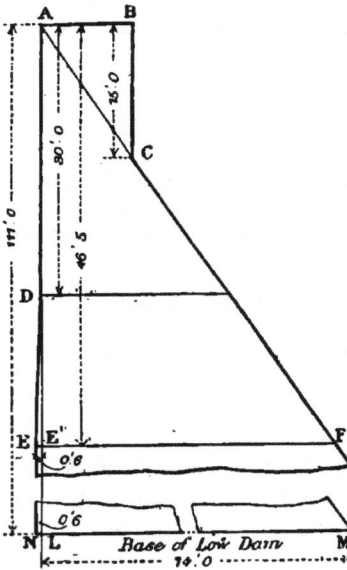

The inner and outer profiles of the dam are then formed by joining the points thus formed—A D E N, A B C M, respectively.

The total area of the section is easily found to be 4225·7 square feet ; and the total weight upon the base N M, including that of the water above the portion D E of the inner profile, may be taken as 265 tons.

242. Proceeding further with our illustration, we may apply the formulæ of § 234 to the investigation of a high dam, assuming the same data as those of the low dam already developed. As these calculations are only intended to illustrate the subject, it will be sufficient here to ascertain points in the profiles at four levels, down to that where $h = 150$ feet below the crest.

The successive values of h, then, under consideration may be thus designated :

From crest to base of low dam, $h_0 = 111$ feet.

,,	,,	1st lamina,	$h_1 = 120$	,,
,,	,,	2nd ,,	$h_2 = 130$	,,
,,	,,	3rd ,,	$h_3 = 140$	,,
,,	,,	4th ,,	$h_4 = 150$	,,

. Corresponding with these values of h, are values of b and x, which are respectively the total breadth of base and the breadth of that portion of it which lies under the inner profile of each successive lamina considered. These quantities, corresponding with h_0, h_1, h_2 . . . ., may be conveniently designated by b_0, b_1, b_2 . . . ., x_0, x_1, x_2 . . . .; and, by a similar notation, the total weight of the vertical slice of the dam and the superincumbent water, corresponding with the successive values of h, may be written W_0, W_1, W_2 . . . .

1st lamina. At the base of the low dam (§ 241)

$$W_0 = 265 \text{ tons, } h_0 = 111 \text{ feet, } b_0 = 74 \cdot 6 \text{ feet, } x_0 = 0.$$

If the profiles above b_0 be produced down to the level $h_1 = 120$ feet, the weight of the trapezoid of masonry thus added is

$$\frac{9 \times 74 \cdot 6 \left(1 + \dfrac{120}{111}\right)}{2 \times 16} \text{, say 44 tons.}$$

Then, as a first approximation,

$$W_1 = 265 + 44 = 309 \text{ tons.}$$

By equation (i.), § 234, introducing the proper value of s,

$$b_1 = \sqrt{\frac{(120)^3}{360}\left(1 + \frac{(120)^4}{5184\,(309)^2}\right)}$$
$$= 82 \cdot 5 \text{ feet.}$$

The next step is to find x_1 corresponding with this value of b_1.

By equation (ii.), § 234,

$$\frac{9}{16 \times 24}[3 \times 5565 - 6806 + 6\,(157 \cdot 1)\,x_1 + 2 \times 6154]$$

$$- \frac{x_1}{36 \times 12}[231\,(165 - 3\,x_1)] - 265\,(2 \cdot 63 - x_1) = 0;$$

that is $\qquad 1 \cdot 6 x_1{}^2 + 199 x_1 - 177 = 0 ;$

therefore $\qquad x_1 = \dfrac{-199 + \sqrt{39,601 + 1133}}{3 \cdot 2} ,$

$$= \dfrac{3}{3 \cdot 2} = 0 \cdot 9 \text{ foot.}$$

The values of b_1 and x_1 thus found, enable us to obtain a closer approximation to the weight of the trapezoid under consideration, and to determine the weight of water over-lying its inner face : The sum of these two quantities may be found to be 47 tons.

Then a second approximation gives

$$W_1 = 265 + 47 = 312 \text{ tons.}$$

Introducing this value of W into equation (i.), § 234, we find, as a second approximation,

$$b_1 = 82 \cdot 3 \text{ feet} ;$$

which value of b_1 is so nearly that previously used in apply-ing equation (ii.), § 234, as not to affect the already found value $x_1 = 0 \cdot 9$ foot.

Thus we have at the base of the 1st lamina below the low dam,

$$W_1 = 312 \text{ tons}, \; h_1 = 120 \text{ feet}, \; b_1 = 82 \cdot 3 \text{ feet}, \; x_1 = 0 \cdot 9 \text{ foot.}$$

2nd lamina. Proceeding as before, produce the profiles immediately above b_1 down to the level $h_2 = 130$ feet ; the weight of the trapezoid of masonry thus added, together with that of the water overlying its inner face, is

$$\frac{10}{16} \left[82 \cdot 3 + \frac{1}{2} \cdot \frac{10}{9} \left(82 \cdot 3 - 74 \cdot 6 \right) \right] + \frac{120 + 130}{2 \times 36} \times 1 \cdot 0 = 58 \text{ tons.}$$

Then, as a first approximation,

$$W_2 = 312 + 58 = 370 \text{ tons.}$$

Hence $$b_2 = \sqrt{\frac{(130)^3}{360}\left(1 + \frac{(130)^4}{5184\,(370)^2}\right)} \quad \ldots \ldots,$$

$$= 92\cdot5 \text{ feet.}$$

Applying equation (ii.), § 234, and introducing these values of b_2 and W_2, we find

$$x_2 = 1\cdot4 \text{ feet.}$$

Correcting the weight of the trapezoid and its super-incumbent water for this value of x_2, we have as a second approximation

$$W_2 = 312 + 59 = 371 \text{ tons,}$$

whence, a reapplication of equation (i.), § 234, gives the corrected value

$$b_2 = 92\cdot4 \text{ feet.}$$

Thus, at the base of the 2nd lamina,
$W_2 = 371$ tons, $h_2 = 130$ feet, $b_2 = 92\cdot4$ feet, $x_2 = 1\cdot4$ foot.

3rd lamina. Produce the profiles above b_2 down to the level $h_3 = 140$ feet.

As a first approximation

$$W_3 = 371 + 67 = 438 \text{ tons.}$$

$$b_3 = \sqrt{\frac{(140)^3}{360}\left(1 + \frac{(140)^4}{5184\,(438)^2}\right)},$$

$$= 102\cdot8 \text{ feet.}$$

By equation (ii.), § 234,

$$x_3 = 1\cdot4 \text{ foot.}$$

This introduces no sensible correction in the value of W_3; and we therefore have, without further calculation, at the base of the 3rd lamina,

$$W_3 = 438 \text{ tons}, h_3 = 140 \text{ feet}, b_3 = 102\cdot8 \text{ feet}, x_3 = 1\cdot4 \text{ foot.}$$

4th lamina. Produce the profiles above b_3 down to the level $h_4 = 150$ feet.

As a first approximation,

$$W_4 = 438 + 75 = 513 \text{ tons.}$$

$$b_4 = \sqrt{\frac{(150)^3}{360}\left(1 + \frac{(150)^4}{5184\,(513)^2}\right)},$$

$$= 113\cdot3 \text{ feet.}$$

By equation (ii.), § 234,

$$x_4 = 1\cdot4 \text{ foot.}$$

Again, no sensible correction is introduced into W_4; and we have at the base of the 4th lamina,

$$W_4 = 513 \text{ tons}, h_4 = 150 \text{ feet}, b_4 = 113\cdot3 \text{ feet}, x_4 = 1\cdot4 \text{ foot.}$$

Summarising these particulars we have :—

Depth below Crest of Dam (h).	Breadth of Base (b).	Weight of Masonry.	Breadth measured from axis under inner face (x).	Weight of Water over inner face.
feet	feet	tons	feet	tons
0	10	..	..	..
111*	74·6	264	0·6	¼
120	82·3	308	1·5	4
130	92·4	362	2·9	9
140	102·8	424	4·3	14
150	113·3	493	5·7	20

The section thus calculated is plotted in *Fig. 47.*

* Base of low dam, § 241.

243. The situations in which most dams are constructed render it necessary for some, often a considerable, portion of the structures to be built below the ground-level. In such cases there must be considered the water-pressure or

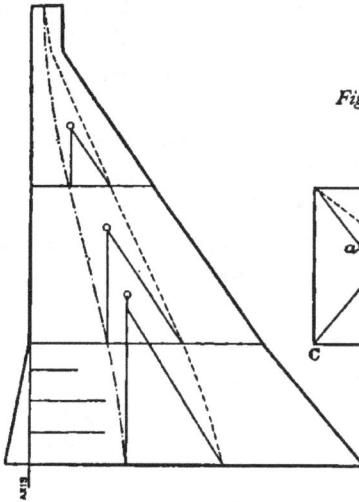
Fig. 47.

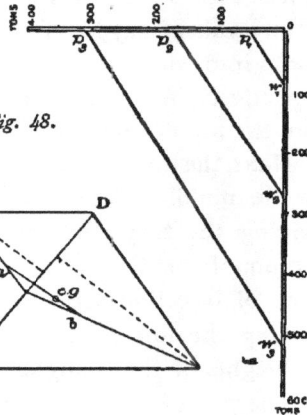
Fig. 49.

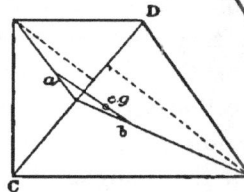
Fig. 48.

Scale, 1 inch = 60 feet.
Line of resultant pressures, reservoir empty · — · — · — · — ·
,,   ,,   reservoir full · · · · · · · · · · · ·

the earth-pressure, according as the ground is, or is not, saturated, acting upon the lower portion of the outer face of the dam, reservoir full; and similar pressures upon the lower portions of both faces, reservoir empty. Both these effects tend to throw the lines of resultant pressures towards the middle of the section. The stability of the dam is thus improved by an amount that may be readily calculated in any particular case, according to the methods discussed in §§ 203 and 234; it is, however, not usual in practice to take advantage of this circumstance to reduce the sections.

244. To illustrate the process of investigating graphically the stability of these structures, the diagrams *Figs. 48* and *49* are added.

The first proceeding is to separate the section into three blocks—in this (illustrative) case by horizontal lines at 60 feet and 111 feet below the top, but in ordinary practice at intervals of, say, 10 feet in depth. The weights of the blocks into which the slice of the dam is thus divided, are, respectively, 81 tons, 184 tons and 248 tons—proceeding from the top downwards.

Next, the *c.g.* of each block is found. *Fig. 48* illustrates, for the middle block, the method adopted: It consists of dividing the trapezoid into two triangles, bisecting their common base, C D, joining the opposite angles with the point of bisection, trisecting the joining lines at *a*, *b*, and dividing the line *a b* at the point *c g* in the inverse ratio of the heights of the triangles above the base C D.

The point *c g* thus found is the centre of gravity of the trapezoid in question ; and the centre of gravity of any portion of the section of the dam is readily found by successive applications of the method—dividing the line that joins the centres of gravity of any two component parts, in the inverse ratio of the areas of those parts.

In the diagram of forces, *Fig. 49*, $o\,w_1$, $w_1\,w_2$, $w_2\,w_3$, are set off vertically to represent the weights of the 3 blocks together with the water over the inner profile of each of them ; $o\,p_1$, $p_1\,p_2$, $p_2\,p_3$, are set off from *o* representing to the same scale the horizontal thrusts of the water—50 tons, 121 tons and 142 tons, respectively, acting against the inner faces of the several blocks.

The lines $p_1\,w_1$, $p_2\,w_2$, $p_3\,w_3$, represent, then, in magnitude and direction, the resultant pressures in the dam acting upon the bases of the blocks. Verticals are now drawn through the centres of gravity of the entire cross-sections corresponding with the several values of *h*—60 feet, 111 feet and 150 feet ; also, from the points situated in these verticals at heights $\frac{h}{3}$ above the corresponding bases, lines are drawn parallel to $p_1\,w_1, p_2\,w_2, p_3\,w_3$, respectively, to cut these bases.

The points in which these two sets of lines cut the bases, if joined as indicated by the dotted lines in *Fig. 47*, form the lines of resultant pressures, "reservoir full," and "reservoir empty."

It may be seen that, in this case, neither of the lines of resultant pressure falls outside of the middle third of the section ; nor do the resultant pressures cause a stress exceeding 10 tons per square foot at either face. Hence, the conditions of strength and stability are satisfactorily complied with ; whilst the close adherence of the lines of resultant pressures to the boundaries of the middle third of the section, indicates that its form is economical in respect of the material of which it may be supposed to be constructed.

245. It will not have escaped notice that investigations relating to the stability of dams, must be affected by the elastic properties of the masonry of which they are constructed. For the purpose of applying such processes of calculation as are practicable, to this case, we have assumed a high degree of rigidity in the structure. We do not propose for acceptance the view that such rigidity is to be looked for in fact, in this or in other materials employed in engineering construction.

The object of the investigation is to arrive simply at an accurate determination of a dam-section, under certain premised conditions, which accord with those obtaining in existing structures of acknowledged strength. Whether, as is assumed sometimes, the plasticity of the materials of which dams are constructed leads to a yielding of the outer portions of their bases, and a consequent occurrence of

maximum stress at some point considerably removed therefrom ; or, as may be imagined from the analogy presented
by the base-pressures of aggregations of granular materials,
the outer portions sustain by an arch-like action the bulk
of the superincumbent mass, and the resulting pressures
diminish towards the middle — extended experimental
knowledge of the internal stresses occurring in masonry,
alone can form the basis of new laws and lead to advances
in practice affecting the design of these structures.

Arched Dams.

246. The examples of dams which depend for their
stability upon the transmission of the water-thrust against
their convex faces to the sides of the valleys which form
their abutments are not numerous.

Although many existing dams are curved in plan,
there are, indeed, only three such structures of any considerable magnitude having a section insufficient to satisfy
the primary conditions of stability, without taking into
account the effect of their curvature upon the distribution
of the stresses transmitted through them.   These are : the
Zola dam,* near Aix, in Provence ; the Sweetwater dam,†
in San Diego, California ; and the Bear Valley dam,‡ in
San Bernardino, California.

The dimensions of the Californian dams are worthy of
note :

	Date.	Height.	Radius at Top.	Breadth at Top.	Breadth at Base.
		feet	feet	feet	feet
Sweetwater dam ..	1888	90	222	12	46
Bear Valley dam ..	1885	64	335	2½	22

* Ann. des Ponts et Chaussées, 1872, tome 3, p. 456.
† Trans. Am. Soc. C.E., vol. xix. pp. 201 *et seq.*
‡ Ibid., pp. 221 *et seq.* Wilson, ' Irrigation Engineering,' p. 264—New
York, 1893.

The latter is built of coursed rubble masonry, and we reproduce its central vertical section in *Figs. 50.*

Notwithstanding the fact that this type of dam is generally regarded with some disfavour, we propose to investigate it briefly ; because, under some circumstances, it may be adopted with decided advantage.

247. Assuming that the thrust of the water, normal to the inner (convex) face of an arched dam, is, in every horizontal lamina of the structure, transmitted to the abutments,

Figs. 50.

SECTION

PLAN

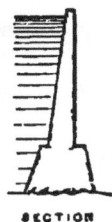

BEAR VALLEY DAM.
Scale 1 inch = 60 feet.

Figs. 51.

SECTION ON AA.

A

PLAN

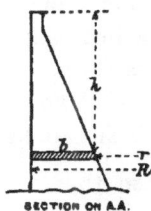

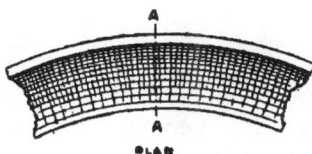

TYPE OF ARCHED DAM.

according to the principle of action of an arch ; the question is freed from the composition of that thrust with the vertical pressure due to the weight of the masonry.

Consider any horizontal lamina of unit thickness, situated at a depth h below the crest, *Figs. 51* (assuming, as before, hydrostatic pressure throughout the entire depth) ; then the thrust transmitted through the lamina is everywhere perpendicular to the normal pressure of the water ; and the

curve of thrust, and therefore the structure in plan, must be circular.

Let R denote the radius of the inner face of the dam, and let w denote the weight of unit volume of water : then the horizontal thrust is, at every point of the curve, $R\,h\,w$.

The breadth of the lamina under consideration must be such as to transmit this thrust without exceeding the maximum permissible intensity of stress in the masonry in any part of the cross-section of the lamina. A difficulty arises here from our imperfect knowledge of the internal condition of an arch under stress ; but the following method of treatment may be employed to obtain results suitable for practical application, based upon the hypotheses as to stress stated in § 222.

Let r_h denote the radius of the outer face of the lamina, which is situated at a depth h below the crest :

Then the ratio $\dfrac{\text{maximum stress-intensity}}{\text{average stress-intensity}}$ in the cross-section of the lamina is

$$\frac{2\,R}{R + r_h} \; ;^*$$

hence, if s_1 denotes the average intensity of stress in the cross-section, whilst s, as before, denotes the maximum permissible stress :

$$s_1 = s\,\frac{R + r_h}{2\,R}.$$

The breadth of the lamina required to transmit the stress $R\,h\,w$ is

$$\frac{R\,h\,w}{s_1},$$

---

* Rankine, ' Applied Mechanics,' 9th edition, Art. 273.

which, by substituting the above value of s_1, may be written

$$\frac{2\,R^2\,h\,w}{s\,(R + r_h)}.$$

But the breadth b (*Fig. 51*) of the lamina is

$$R - r_h ;$$

hence

$$R - r_h = \frac{2\,R^2\,h\,w}{s\,(R + r_h)} ;$$

therefore

$$r_h = R\sqrt{1 - \frac{2\,h\,w}{s}} ,$$

and

$$b = R - r_h = R\left(1 - \sqrt{1 - \frac{2\,h\,w}{s}}\right). \qquad \text{(i.)}$$

248. Equation (i.), § 247, gives the necessary breadth of the dam at any given depth h below the crest. This varies directly as R, and, in ordinary practice, will be found to be less than that required for a gravity dam, only when R is less than 250 feet—provided that the same maximum stress-intensity is adopted in both cases. Also, as may be seen by inspection of the equation, the breadth increases with the depth, until, when $h = \dfrac{s}{2\,w}$, it becomes equal to the radius, and the arched dam is impossible.

It would appear, therefore, that the use of this type of structure must be very limited, unless a higher intensity of stress be permitted than that ordinarily contemplated in the design of dams. And this is found to be the case in practice. There are grounds for considering that the limiting stress may, under certain special circumstances, be placed higher than is usual in ordinary masonry structures.

R

In the case of the Bear Valley dam alluded to above, it may be observed that, writing equation (i.), § 247,

$$s = \frac{2\,h\,w}{\dfrac{b}{R}\left(2 - \dfrac{b}{R}\right)}$$

we find that at 48 feet below the crest, where the breadth of the dam is 8·5 feet, the maximum stress in the cross-section is, according to the dimensions stated,

$$\frac{2 \times 48 \times \dfrac{1}{36}}{\dfrac{8\cdot5}{335}\left(2 - \dfrac{8\cdot5}{335}\right)} = 53 \text{ tons per square foot.}$$

249. It must not be overlooked that fracture of the wall from any accidental cause would probably be attended by immediate and total collapse ; and any dams exhibiting such a high stress-intensity must be regarded as, in any event, unsuitable for situations where the sudden release of the waters impounded by them would endanger life or valuable property.

For mining, irrigation, and other industrial works, where the economy resulting from the adoption of this type of dam is frequently an all-important consideration, it is deserving of attention.

250. A third class of masonry dams may be noticed under the designation of " composite dams," that is to say, walls which, possessing sufficient strength and stability to act as gravity dams, are curved in plan with a view to endow them with a further factor of strength from their action as horizontal arches.

Although numerous examples of this class are found in practice, it is, when the dams are large, one that does not merit commendation.   Generally speaking, if the radius of the inner face exceeds 500 feet, the operation of arch-thrust would give rise to stresses in the masonry which are not warranted by experience.

The prolonged existence of those examples of arched dams of large radius must be ascribed to the fact that, owing to the elastic nature of the material used, the arch does not act at all—or acts in some unknown degree, producing slighter stresses than present theories would lead us to expect ; or else the dams are existing in a condition of internal strain never contemplated by their designers.

251. Turning to existing examples of gravity dams, we find that, notwithstanding the attention they have received from engineers during 300 years, these structures exhibit a variety of practice so great as to be inexplicable, except on the supposition that the mechanics of the subject have been often imperfectly realised. To those who desire to refer to the history of the question, we commend the treatise of Mr. Edward Wegmann,* in which may be found diagrams and other information relating to nearly 40 dams —including the famous dam of Furens, which, notwithstanding the elegance of its proportions, imposes a maximum stress of only 6 tons per square foot upon its base ; and the more recent Gileppe dam, the section of which exhibits a mass of masonry more than double that of the former. As examples of more recent practice, *Fig. 52* shows the section of the Vyrnwy dam of the Liverpool Waterworks, completed in 1889, and *Fig. 53* illustrates the Tansa dam of the Bombay Waterworks, completed in 1891. The former of these, built of " Cyclopean rubble " masonry, in Portland cement, develops a maximum intensity of stress amounting to 8·7 tons per square foot ; although it should be observed that, under the conditions ordinarily prevailing,

* Wegmann, 'The Design and Construction of Masonry Dams'—New York, 1888.

the maximum stress-intensity does not exceed 6·6 tons per square foot.*

The Tansa dam, *Fig. 53*, is constructed of rubble masonry in hydraulic lime. Its maximum stress-intensity is 7·5 tons per square foot—occurring in the outer toe, reservoir full. This dam does not strictly comply with the primary condition (2); inasmuch as the line of resultant pressures, reservoir empty, falls outside the middle third of

Fig. 52.

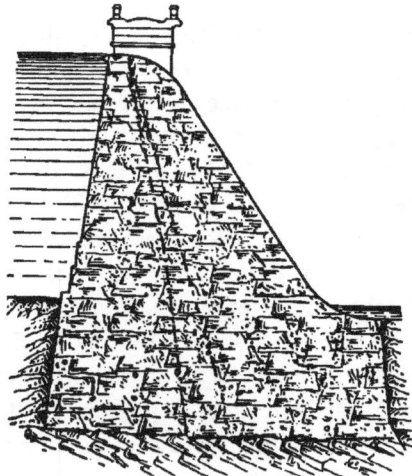

VYRNWY DAM. Scale 1 inch = 60 feet.

the section between the base and 60 feet below the crest— whence may be inferred the occurrence of tensile stress in the outer face of that portion of the dam. In section it occupies a mean position between that of the Vyrnwy dam and the Stony Creek dam of the Geelong Waterworks, erected in 1873, *Fig. 54*. This last-mentioned work is built of concrete; it is arched in plan with a radius of

* 'Report of Mr. G. F. Deacon as to the Vyrnwy Dam'—Liverpool, 1885.

300 feet at the crest; the top-breadth is only 2½ feet, although the structure itself, like that at the Vyrnwy, performs the office of waste-weir, being coped with heavy stones.*

252. The Stony Creek dam is, however, only about one-half the height of the Tansa dam—the length of exposure or "fetch" being about half a mile. In both these particulars it much resembles the dam of the Abbeystead (compensation) reservoir of the Lancaster Waterworks,† com-

Fig. 53.

Fig. 54.

STONY CREEK DAM.
Scale 1 inch = 60 feet.

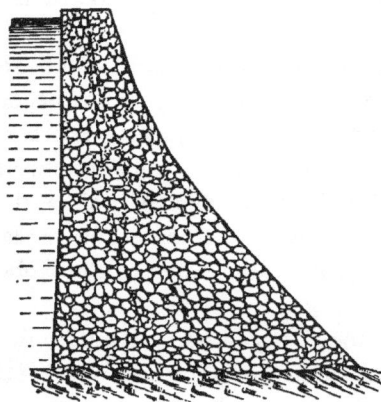

TANSA DAM. Scale 1 inch = 60 feet.

pleted in 1880 by Mr. James Mansergh. The Abbeystead dam affords an early, if not the earliest, example of this kind of work erected in Great Britain. Its top-breadth is 7½ feet; and it may be observed that, in the account of the Stony Creek dam referred to, the author of the paper prudently suggests that a greater top-breadth than 2½-feet would be desirable in other similar structures.

Attention is directed to this matter, as there is at

* Minutes of Proceedings Inst. C.E., vol. lvi. pp. 113 and 114.
† Ibid., vol. lxviii. pp. 263 *et seq.*

present no generally accepted rule for the top-breadth of a dam. It has been sometimes proposed that this feature should vary with the height of the structure ;* and cases have occurred where it has been settled merely by appearances. It should be governed by the local circumstances of "fetch," and the road-way requirements. Its importance is more considerable than might at first sight appear, because certain notions of symmetry, which frequently form an element in the design of these structures, are affected by the dimensions assigned to the feature in question, which thus indirectly governs the proportions of the entire work.

253. The Tytam dam at Hong Kong,† built of concrete with ashlar faces, is 120 feet high, and presents the uncommon feature of a stepped outer profile, the steps being 10 feet in height. This dam and the aqueduct works in connection with it, visited by us shortly before their completion, form an ornament to the colony, and present some important constructional features due to Sir Robert Rawlinson.

254. The section of the Periyar dam, Madras, illustrated in *Fig. 55*, is interesting, as affording one of the most recent examples of design in this class of work, and showing a peculiarity in the form of the upper portion, which it shares with a few other dams, notably the Furens. It is now in course of construction in hydraulic lime concrete, gauged five of aggregate to one of the matrix, under circumstances of exceptional difficulty in regard to the heavy floods in the river that traverses its site. The limiting stress-intensity proposed in the design is stated to have been 8 tons per square foot.‡ Here, again, the primary

* Minutes of Proceedings Inst. C.E., vol. lxxxv. p. 282.

† Ibid., vol. c. p. 248.

‡ *Engineering*, vol. liv. p. 654 ; Buckley, ' Irrigation Works in India,' London, 1893.

conditions of strength and stability are not adhered to ; and, in the case " reservoir empty," the resultant weight of the dam falls as much as two feet beyond the middle third of the base, thus inducing tensile stress in the masonry at

Fig. 55.

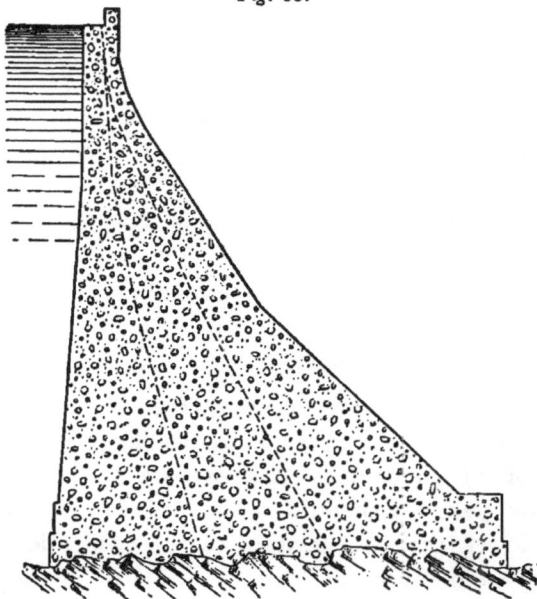

Periyar Dam.   Scale 1 inch = 60 feet.

the lower part of the outer face.   A slightly different disposition of the material would have obviated this, without increase of section.

255. In this connection we may mention that the empirical formulæ for determining the profiles of masonry dams, proposed in 1874 by Sir Guilford L. Molesworth,* are the only concise equations with which we are acquainted applicable to this subject, involving rigid compliance with

* Roorkee Professional Papers, Nov. 1874, pp. 394 *et seq.*

the primary conditions of stability. The element of the specific gravity of the masonry as a factor of the expressions is, however, omitted from them. These equations are, linear measurements being in feet :

$$y = 7\cdot5.\sqrt{\frac{x^3}{P}},\qquad\text{(i.)}$$

and
$$z = \frac{y}{10},\qquad\text{(ii.)}$$

Where x denotes depths below the top of the dam ;

y denotes ordinates to the outer face from a vertical through the crest ;

z denotes ordinates to the inner face from the same vertical ;

and P denotes the maximum stress-intensity in lbs. per square foot.

y has, as a minimum, the value $0\cdot6\,x$.

The value of P, above, applies to the vertical component of the actual resultant pressure. The difficulties to which the use of this component gives rise, are met in a certain degree by gradually diminishing the values assigned to P in the lower part of the dam, as the inclination of the resultant to the vertical increases. Thus, for very high dams, a decrement of the limiting stress-intensity is introduced by writing equation (i.)

$$y = 7\cdot5\sqrt{\frac{x^3}{P(1 - 0\cdot0013\,x)}}.$$

A rule is also given for finding the top-breadth of a dam, thus :

Let h denote the total height of the dam ;

β denote the value of y at the base ;

and a denote the value of y at $\dfrac{h}{4}$ below the top ;

then the proper top-breadth is proposed to be

$$a - \frac{\beta}{40}.*$$

256. The construction of the vast masses of masonry or concrete of which dams are formed, is an art that requires the exercise of judgment in the selection of the materials used, close attention to their preparation, and watchfulness during the process of building. The question as to whether concrete or rubble masonry is, in any given case, the preferable building material, must largely depend upon the character of the rock available. The great tensile strength of Portland cement causes it to be in high favour as a matrix; but it is not entirely unexceptionable. Where a dam abuts, as is frequently the case, upon steep hill-sides, the rapid variation in height tends to produce an unequal settlement in the masonry or concrete as the structure is gradually raised. This settlement is resisted by the shear of the material transmitted from the comparatively un-yielding rock abutments. It is not improbable that action of this kind has produced some of the fractures that have too often occurred in such dams. If this be so, it points to the prudence of building in a slow-setting hydraulic lime, thereby permitting free settlement of the mass before it sets rigidly. Otherwise it would appear that the dam should be so designed that the intensity of the stresses in high sections may be less than that at the same levels in lower sections of the structure; and consequently, that the strain at all parts may tend to vary inversely as the height of the dam at those places.

---

* By introducing the values of a and β, as determined by equation (i.), into this expression, it reduces to $\frac{\beta}{10}$.

It should be noticed that dams in India may be assumed to be free from the thrust of ice against their crests, a factor which in United States practice possesses considerable importance.

The rapid settlement during construction of large masses of concrete, which may be regarded as bearing, beam-like, upon their extremities, must give rise to conditions of internal shearing-stress of indeterminate amount. In concrete work, such settlement is much reduced by the introduction of large blocks of stone and boulders into the work—a practice that possesses the advantages of adding to the coherence of the entire work, and, generally, of effecting some economy in its cost.

It is highly important for the work to be homogeneous in composition and free from plane joints. Similar precautions to those used in the case of earthen embankments, must be taken to avoid the sealing of springs in the foundations ; and facility of escape at the outer face of the dam should be afforded to water percolating the mass.

257. The building of rubble masonry is an art that is hardly amenable to literary treatment. The all-important subject of the manufacture and laying of concrete in engineering structures for hydraulic works, is too wide to be fully dealt with here ; and we abstain from presenting an inadequate description of the numerous manipulative processes necessary for the production of good concrete-work, believing that a serviceable knowledge of the matter can be best gained by observation. The most important of the points to be observed are : the cement is to be thoroughly slaked by aëration before use ; the material composing the aggregate is to be clean, and of such dimensions that it may form, with the cement, a compact conglomerate : the concrete is not to be re-assorted after the completion of the mixing process, as happens if it is either tipped into place from too great an elevation, or is unduly "punned" or beaten with a view to consolidate it ; the quantity of water added in mixing the concrete is to be not

only that required for the hydration of the cement, but, in addition, sufficient to meet the loss by evaporation and by absorption into the materials and the adjacent bodies. The last-mentioned point is, perhaps, the most difficult of observance. It is undoubtedly improper to over-water concrete ; but at the same time, the practice of the testing-room must not be carried on to works, where a different set of conditions obtains. To obtain successful results, the ratio of water to cement must, in the manufacture of concrete for hydraulic work, be considerably higher than that employed in the production of cement briquettes of high tensile resistance. The same remark applies to cement mortar for masonry and brickwork. Finally, a condition of perfect quiescence must be ensured, in order that a mass of concrete may set well. In foundation work especially, it is far better for concrete to be actually laid in water, than for the water to be drawn away through or underneath the mass during the process of setting, as is sometimes done by heavy pumping. Although the best results may be obtained from concrete set in air, the damage wrought by the percolation of water through it before it has set, and the consequent removal of part of the cement, is irreparable.

258. After all possible care has been exercised in the selection and the mixture of the materials that compose mortar, and in the laying of masonry, the question as to what stress it is permissible to impose upon a structure built of these materials is still far from being satisfactorily answered. The pressure required to crush small blocks of good concrete may be found to be as much as 150 tons per square foot, and it might be thence inferred that a stress-intensity of 25 tons per square foot could be safely assigned to a mass of that material. On the other hand, experiment points to some diminution of the resistance to crushing offered by similar blocks of mortar and concrete, according

as their diameters increase ; * and it is probable that, as a rule, small selected samples of materials are comparatively so free from defects of composition as to yield relatively higher results under stress than do large samples of similar figure. This, it is well known, was the origin of the fictitious strengths attributed by the earlier experimenters to wooden beams—their results being derived from tests of small, straight-grained, seasoned bars of timber ; and caution must be used in applying to practical cases the results of laboratory crushing-tests of small cubes of stone and concrete, especially when they fail to indicate the point at which deformation of the samples first occurs.

Some valuable determinations of the resistance to crushing of concrete prisms cut out of the Vyrnwy dam, were made by Messrs. David Kircaldy and Son in 1885, the lowest result obtained being 181 tons per square foot.† Among walls for impounding water, the Almanza dam, in Albacete, Spain, a structure of ancient date, bears a calculated intensity of stress equal to 12 tons per square foot ; but a lower stress is observed in most modern dams. Still, with good materials and workmanship, there does not seem to be any reason why this, or even a somewhat higher stress should not be adopted with perfect safety.

259. In two respects a masonry dam possesses a distinct advantage over an earthen embankment : (1) It can form its own outlet, (2) the waste-weir works may be safely carried through the body of the structure. The dimensions of both the outlet and the waste-weir are determined by the same considerations as those treated in §§ 199, 216.

The top of the dam may be rounded to form the waste-

* Gillmore, 'Notes on the Compressive Resistance of Freestones, &c.' —New York, 1888.

† 'Report on the Vyrnwy Masonry Dam,' by Major-General Sir Andrew Clarke—Liverpool, 1886.

weir, like that of the Vyrnwy, a proper apron being formed at the outer toe, to receive and check the velocity of the overflow ; or the weirs may be formed at the ends, supplemented by stepped waste-water courses carried down the hill-sides.

The scour-valves may be placed in a culvert through the lower part of the dam, and be manipulated from a chamber overhead, and the draw-off valves may be situated in a stand-pipe or tower within the reservoir.

260. Among the important details connected with the completion of every large reservoir, is the protection of the banks in the neighbourhood of the (varying) water-line, from the erosive action of waves, by stone-pitching or by mattresses. Another important point is, especially in the case of reservoirs formed in valleys into which the tributary streams discharge with high velocity, to cut off such tributaries by means of rubble or timber screens, above the level of the reservoir, so as to form catch-pits to intercept the detritus borne down by torrents, and to prevent it from entering the reservoir, to the diminution of its capacity and the detriment of the impounded water. Such catch-pits are cleaned out from time to time, the intercepted material being formed into a spoil-bank in some convenient and out-of-the-way situation.

SERVICE-RESERVOIRS.

261. The requisite capacity of a service-reservoir depends upon the variation in the rate of daily consumption of water in the district supplied through the reservoir, and upon the length of time during which it is deemed prudent that the water-service should be independent of the supply from the main source, to meet cases of accident, failure or abnormal draught (§§ 190–192).

Elevated situations in the immediate vicinity of the districts to be supplied, are naturally chosen, if such occur and are available for the purpose, as sites for service-reservoirs. It was formerly considered unnecessary to roof these reservoirs, which were often constructed as open pits, half sunk and half embanked, and pitched with stone over their inner slopes. Many examples of this kind remain in use at the present day. For the supply of water to towns, however, experience has taught the superiority of covered reservoirs—excluding dust, and preventing the development of vegetable and animal life in the water, which is due to the action of heat and light.

The subject of the present practice in regard to the construction of service-reservoirs has been so exhaustively treated in an article published by the Institution of Civil Engineers,* that we cannot do better than refer the reader to that paper, and to the subsequent discussion thereupon.

262. The ordinary type consists of concrete walls and floor, backed by 1 foot to 3 feet of puddled clay according to the depth of the reservoir ; the walls, where they rise above the level of the natural ground, being supported by embankments of earth.

The earthen backing behind the walls, and the thrust of the floor and roof inside the reservoir, permit the employment of sections considerably lighter than would be required to resist the hydrostatic pressure of the full reservoir, or the unbalanced external pressure, acting upon simple walls ; but in retentive soils, where heavy pressures may be expected, the walls and floor must be duly increased in thickness—a system of drains being first laid underneath the floor, so as to prevent the accumulation of land-water there, and consequent "up-lift." A series of brick or con-

* " Covered Service-Reservoirs," Minutes of Proceedings Inst. C.E., vol. lxxiii.

crete arches carried on brick or iron pillars springing from
the floor, support an earthen covering some 2 feet or 3 feet
in thickness, *Fig. 56*.

Fig. 56.

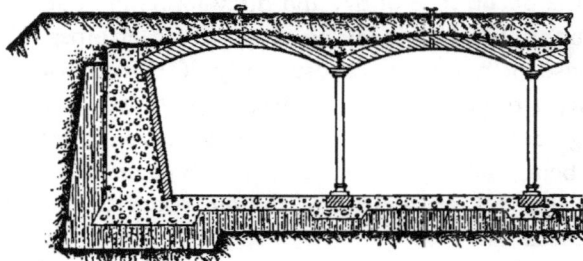

It need scarcely be said that every service-reservoir or
tank should, in addition to inlet- and outlet-pipes (the
former provided with an automatic cut-off valve, and if
necessary with a reflux-valve (§ 355), be furnished with a
scour-pipe and with an overflow having a discharging
capacity at least equal to the delivery of the inlet-pipe.

263. As a detail in the execution of such works, it
should be observed that where a heavy wall united to a
broad floor is built upon puddle or other plastic material, it
is desirable to build the walls to a considerable height
before joining the floor to them. By this plan, the risk of
fracture of the floor is reduced. Where puddle is used,
all steps or footings at the back of a wall should be bevelled
so that, in settling, the puddle may slide off them and not
crack at the angle.

Fig. 56 shows a section of a service-reservoir wall of a
type which has been used by us with satisfactory results.
The puddled floor is prepared for the reception of the
concrete by a paving of large shingle or similar material
pressed into the soft surface as soon as the requisite thickness is laid. The various precautions mentioned in § 212

must be observed in the preparation of the puddle ; and where it is carried up in a thin sheet behind a wall, it is desirable to arrange the work so as to lay it as continuously as possible—thus avoiding many of the failures that occur from the occurrence of dry and dirty joints in puddle-walls. An important point in laying scour- or outlet-pipes through the base of a high wall, especially if this is backed by puddle, is to pass the pipes through ports left in the walls during construction, so that they may be free to move—the ports being subsequently plugged with cement around the pipes after all settlement of puddle, concrete, or masonry has ceased.

In rendering service-reservoirs water-tight, good results have been frequently obtained from an asphalt sheet, $\frac{3}{4}$ inch to 1 inch thick, run whilst hot over the floor, and either into a cavity left in the walls, or laid upon the latter in successive coats by the brush. Precaution must of course be taken to prevent the occurrence of bad joints in the asphalt, which would happen if one coat were laid upon a cold under-layer.

264. Tanks, elevated considerably above the ground, are frequently employed to maintain steadily the pressure in the service-mains and distributing-pipes of a district under supply. Such tanks are generally constructed of iron, and rest upon brick or stone towers, or, more seldom, upon iron columns braced together by lattice - work. Formerly, cast- or wrought-iron tanks of rectangular form, stayed internally, were supported on girders, but more modern structures consist of steel boiler-plate cylinders, resting upon a concrete or brick foundation—the bottom being formed of either metal carried on girders or arches or, in cases where elevation of the tank above the ground-level is unnecessary, brickwork or concrete rendered with asphalt. The principal difficulty attending the construc-

tion of such tanks arises at the junction of the sides and floor—especially when these are made of different materials. Here a rigid joint is inadmissible, owing to the distortion that arises in the tank from changes of temperature ; whilst water-tightness under considerable pressure is often required. A form of joint which has much to recommend it, and has given good results in practice, is illustrated in *Fig. 57.**

Fig. 57.

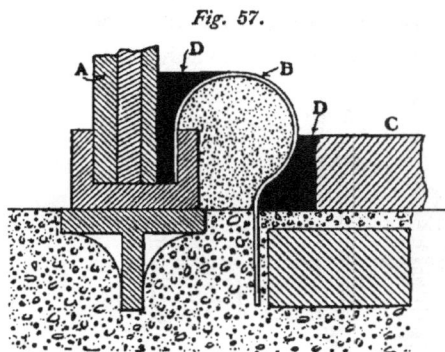

A, shell of tank (iron) ; B, sheet-lead roll ; C, floor of tank (brick) ;
D D, asphalt.

Scale 1 inch = 1 foot.

265. By far the most novel and interesting type of service-reservoir—combining strength, economy and elegance—is presented by the spherical-bottomed tank, which, requiring neither internal stays nor supporting-girders, may be built upon a tower or upon a simple curb supported by columns or piers at any desired elevation. A section of a tank of this kind is shown in *Fig. 58.*

The central shaft contains a ladder or spiral stair, by which access is afforded to the top, and thence to the interior of the tank. The inlet- and overflow-pipes are shown on either side of the shaft, and are provided with

* Proc. Roy. Soc. of N.S.W., vol. xxv. p. 158.

S

expansion-joints, glands and packing, to meet the changes in the sag of the spherical bottom and expansion and contraction of the pipes. The valves are situated at the feet of these pipes. Sometimes, both inlet- and outlet-pipes are provided, to promote circulation of the water through the tank. The roof is double, with a view to secure equability of temperature, and is carried on radial steel principals. In hot climates the shell of the tank may advantageously be surrounded by wood-lagging, a few inches distant from it. The suitability of this form of

Fig. 58.   *Fig. 59.*

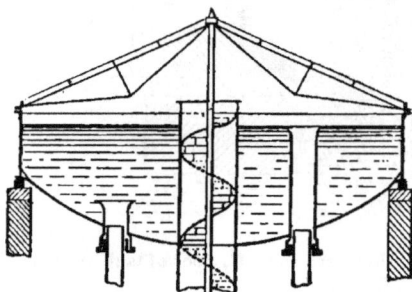
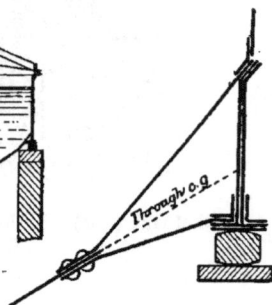

Scale 1 inch = 20 feet.   Scale ⅛ inch = 1 foot.

reservoir for extremely cold climates is questionable. Among examples that are worthy of notice, are that at the Lynn Waterworks and that at Norton in Cheshire, belonging to the Liverpool Waterworks. The former, 26 feet in diameter and containing 55,000 gallons, constructed by Messrs. Easton & Anderson, of Erith, in 1875, is one of the earliest examples of the type in England; the latter, the latest example, is 76 feet in diameter and has a capacity of 600,000 gallons.

The calculations of the strength of this form of tank

are of a simple character. For the thickness of the cylindrical shell, at h feet below the top, we have

$$t = \frac{R\,h\,w}{12.s} + \frac{1}{4} \; ;$$

where t is the required thickness in inches at h feet below the top ;

R is the radius of the tank in feet ;

s is the working-stress of the metal in tons per square inch ;

and w is the weight of 1 cubic foot of water expressed as a fraction of a ton, i. e. $= \frac{1}{36}$.

The thickness for the spherical bottom is arrived at by the application of the well-known theorem that the intensity of the tension in a spherical shell is one-half of that due to equal fluid-pressure within a cylindrical shell of the same radius.[*] Hence, using the same notation and putting r for the radius of the spherical bottom in feet, we have, treating the weight of metal as its equivalent head of water,

$$t = \frac{r\,h\,w}{24.s} + \frac{1}{4} \; .$$

The bottom plates have to be specially shaped in dies, but this introduces no insurmountable difficulty. Riveted lap-joints for the circular seams, and butt-joints and strips for the radial seams, are found to answer well for both the bottom and sides. The interior surface of the tank may be rendered with either Portland cement or asphalt, the seams being first carefully caulked, and the plates freed from dust and dirt. The outside seams should also be carefully set up. One of the most important features of this type of tank is the compression-ring to which the bottom is attached. A form of section which possesses many advantages for this ring is shown in *Fig. 59.*

* Rankine, 'Applied Mechanics,' 9th edition, Art. 272.

S 2

In it the rivets are, as far as possible, in shear, and the attachment of the bottom plates to the compression-ring presents little difficulty. The tangent at the edge of the spherical bottom passes through, if produced, the centre of gravity of the compression-ring, the latter being carried on the cast-iron curb, or bed-plate, by means of short iron rockers, shaped so as to lift the ring slightly in rolling, and tending to regain their normal position—a necessary provision against horizontal movements of the tank due to changes of temperature. The bed-plate must be capable of resisting the thrust so induced.

266. In the United States, it is a common practice to form service-reservoirs of high cylindrical tanks of large diameter. As an example, that at Savannah, in Georgia, is 30 feet in diameter and 57 feet high, constructed of iron boiler-plate, and is carried by iron girders upon a brick tower 50 feet in height.* Many tanks of still greater height rest simply upon a prepared foundation on the ground.

Where such height is necessary, the upper portion only of the tank can be regarded as effective for the general purposes of a service-reservoir, and the lower portion may be conceived to act as a supporting column, operating through the medium of the circumferential tension in the shell, and providing, of course, in particular cases, a useful reserve of low-pressure water. There are doubtless good reasons for the popularity of this design in some localities abroad; but it cannot be proposed as a suitable form of service-reservoir for erection in this country.

267. We will now allude briefly to the subject of the storage of water on a small scale in houses. For manufacturing purposes, convenience and cost determine the position, size and material of internal cisterns and tanks. But when water has to be stored in cisterns, to be subse-

* 'Reports on the Water-power of the United States'—Washington, 1887.

quently used for domestic purposes—which is often the case, even with constant supply ; owing to the excessive draught upon the service-pipes during the busiest hours of the day—the matter demands earnest attention. Too frequently house-cisterns are so large in relation to the domestic consumption, that the water they contain cannot, under the ordinary conditions of use, be entirely changed for many hours, perhaps days. They are commonly situated so near to the roof and the hot-water apparatus, and in a chamber so devoid of facilities for its ventilation, that the water contained in them is never cold when it is required for use ; and, in addition to this, they are often lined with some corrosive and poisonous metal.

Such a system of storing water for domestic purposes cannot fail to be injurious to the health of the consumers ; and, by it, the benefits of a pure and cool water supply; derived perhaps at great cost from some distant source, are, at the very latest stage, often largely annulled.

House-cisterns should be of such dimensions as to hold only 24 hours' supply of water. They should be constructed of slate or of galvanised iron, and should be situated in an airy chamber, not too near the roof, and as far as is convenient from the hot water supply cistern or other heating-apparatus. Finally, they should be provided with ventilating covers and with plugs at the bottom, to permit of their being properly cleansed and scoured out from time to time. The circumstances that affect the purity of water under storage will be more fully considered as part of the general question of purification.

CHAPTER V.

THE PURIFICATION OF WATER.

268. WATER, collected in the condition in which it
occurs in nature, with the exception of that derived from
springs and wells, must always be subjected to some
treatment before it can be pronounced fit for domestic and
industrial use ; and even spring- and well-water, although
generally free from organic pollution, is often impregnated
with mineral solutions that injuriously affect its quality.
The term "purification" here means the operation of
rendering any natural water fit for distribution through
mains and service-pipes to meet the requirements of consumers. This does not invariably mean improvement of
the chemical purity of the water, although such is generally
the case ; in fact, purification, in the sense in which the
waterworks engineer understands it, so far from being restricted to the removal of constituents foreign to chemically
pure water, involves sometimes the addition of substances
to it, in order to improve its dietetic properties, or to
reduce its corrosive action upon the vessels in which it is
contained.

Such a course has been taken, for example, at Dessau
in Anhalt, where the water supply is very soft, containing
much carbonic acid in solution, and was found to dissolve
lead from the interior of the service-pipes in dangerous
quantity. By the addition of powdered carbonate of lime,
the excess of carbonic acid has been removed from the

water, and its solvent action in the formation of lead carbonate prevented.

269. The impurities contained in water occur either as floating on its surface, in suspension, or in solution; and the removal of foreign matter in the first and second of these conditions, and to some extent also in the third, constitutes the purification that water most commonly undergoes.

(1) Floating matter may be removed by means of an overflow from a reservoir or other convenient chamber, whence such bodies are carried away by the escaping water; or they may be strained out of the water by causing it to pass through gratings or wire gauze.

(2) Matter in suspension, if not too minute, may be strained out by wire gauze; or, if of specific gravity higher than unity, it may be permitted to precipitate naturally by ensuring to the water a sufficient period of quiescence. All storage-reservoirs produce this very useful secondary effect upon their contents. Fish and other inhabitants of streams and ponds assist actively to absorb suspended organic matter, and to a certain extent also, that which floats upon the surface. Water that carries silt is clarified by natural subsidence in ponds, or by flow through shallow tanks at a rate slow enough to permit of the deposition of the solids upon the floors. If the silt is in a state of exceedingly fine division, its settlement is hastened by producing an artificial precipitate in the water. Thus, if turbid water contains bicarbonate of lime in solution, the addition of either cake-alum* or lime produces a heavy precipitate that rapidly carries the silt down with it to the bottom. The addition of the former compound causes a chemical action of double decomposition, resulting in the formation of aluminium hydrate which falls as a precipitate,

* Aluminium sulphate, not true alum,

of carbonic acid which escapes, and of calcium sulphate which remains dissolved in the water, thus,

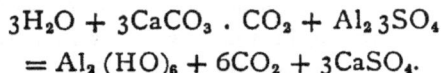

$$3H_2O + 3CaCO_3 \,.\, CO_2 + Al_2\,3SO_4$$
$$= Al_2\,(HO)_6 + 6CO_2 + 3CaSO_4.$$

The action when the latter compound is added is to produce a precipitate of calcium carbonate, thus,

$$CaCO_3 \cdot CO_2 + CaO = 2CaCO_3,$$

as in Clark's " softening " process (§ 307).

In both cases, the sudden formation and settlement of the precipitate carries suspended silt down with it, and exercises a marked clarifying effect upon the turbid water.

After removing by subsidence such matters as readily settle, it is the usual practice to further purify water by filtration, which, in addition to removing turbidity, greatly reduces the quantity of micro-organisms present.

· (3) Water may be freed from some substances in solution by aëration, or exposure to the action of the atmosphere—the rapidity of the operation being greatly promoted by agitation : e. g. if water containing dissolved sulphuretted hydrogen be caused to flow along an open channel the gas escapes ; similar treatment applied to a more common impurity, carbonate of lime in solution, causes carbonic acid to be evolved and carbonate of lime to be deposited as a precipitate in the channel. As another special example, the water-supply of Königsberg[*] is notably rich in iron compounds, chiefly ferrous carbonate dissolved in excess of carbonic acid, which have been found to be favourable to the growth of *crenothrix*—an

[*] Deutsche Vierteljahrsschrift für öffentliche Gesundheitspflege, 1888, p. 339.

organism that is capable of rapid development even in dark situations—advantage was here taken of the fact that the line of the main is intersected by an ancient water-course, into which the water is turned and permitted to flow down for a length of about five miles. The result is that carbonic acid escapes, ferric oxide is deposited in the channel, and the water issues at the lower end clear and bright, its tendency to promote organic growth being altogether lost.

270. On shipboard, and to a limited extent on the sea-coast, in certain places where other sources of supply are wanting or inadequate, as at Aden, distillation of sea-water is resorted to as a means of ridding it of the excessive quantity of salts that render it unfit for domestic use. A good example of this process at Iquique, Peru, has been described by Mr. C. M. Johnson, R.N.[*]

The plant comprises a series of five cylindrical generators, a Root multitubular boiler and a surface-condenser, and is capable of supplying about 45,000 gallons of distilled water per day. There are two nests of wrought-iron tubes running through each generator placed on either side of the centre line. The steam from the boiler, at 140 lbs. pressure per square inch, enters the tubes of the first generator, causing the sea-water in it to evaporate. Part of the steam thus generated is led through the tubes of the second generator, heating the brine in it, and so on throughout the whole series. From the generators the bulk of the vapour is conducted through the surface-condenser into a fresh-water tank. When the brine becomes by this process concentrated to about three times the density of sea-water, it is discharged, part of its heat being abstracted by causing it to circulate about the feed-pipes of the boiler and generators.

[*] Minutes of Proceedings Inst. C.E., vol. lxxvii. p. 344.

The removal of compounds that are truly hydrated, or in chemical combination with water, hardly comes within the province of the engineer.

271. We proceed to consider in more detail the operation of purifying water for domestic and industrial purposes, and to notice the apparatus necessary in each particular case. The several important processes may be summarised as those of settlement, straining, filtration and precipitation. It seldom happens that all of these processes are applied to any one water, but only such as are most applicable to the individual case.

272. When surface-water is collected into a storage-reservoir or river-water is pumped into a tank, the subsidence of matter in suspension, if heavier than water, commences as soon as any portion of the water attains such a state of quiescence that the velocity of its internal whirls and eddies is no longer sufficient to buoy up the material in question. Floating bodies are partially removed by the overflow of the reservoir. As may be naturally inferred, all the conditions that conduce to stillness in the body of the water, favour such clarification; whilst on the other hand, commotion has an advantage in promoting purification by aëration.

As the internal whirls and eddies in the water die out, the suspended matter, if of sensible size, very soon falls to the bottom. But if it be in an extremely fine state of division, so that the ratio of the amount of surface of each individual particle to its volume is large, its downward velocity is very slow, and the resistance to its motion then varies as the velocity instead of with some higher power of it.

That is to say,

$$\frac{dv}{dt} = -kv$$

where v is the downward velocity at the time t and k is a constant.

Then $log \dfrac{v}{V} = - k\,t$, V being the velocity when $t = o$.

Therefore $v = V\,\epsilon^{-k\,t}.$

Integrating, $s = \dfrac{V}{k}\,(1 - \epsilon^{-k\,t}).$

This equation shows that if $t = \infty$, $s = \dfrac{V}{k}$; i. e. in an infinite time such a particle only settles a finite distance equal to $\dfrac{V}{k}$. From this it follows that water containing suspended matter of impalpable fineness would never clear by settlement.

273. The draw-off works of a reservoir or the intake works of a river supply should be placed in such a situation as to command the clearest water. This necessitates their being removed from places where the entrance or confluence of streams causes commotion in the water, or where the nature of the beach is such that a turbid condition is liable to be produced by the action of the waves upon it. Before leaving the reservoir or entering the pump-well, as the case may be, previous to filtration water is passed through strainers of copper or brass wire gauze. The size of the mesh employed varies according to the impurities dealt with. That of the Glasgow Waterworks is $\frac{1}{40}$ inch; that of the Vartry works, Dublin, is $\frac{1}{30}$ inch, whilst that introduced on the new Liverpool Waterworks at Vyrnwy has 100 wires per lineal inch. For cleansing fine gauze strainers, some special process must be applied, such as the action of jets of water impinging on their inner surface. The straining-chamber

is constructed either in the reservoir or river as a tower, or as a well in the adjacent ground. It is frequently circular in plan, the strainers being arranged in polygonal form concentric with the well; the inlet then communicates with the annular space thus formed, and the clear-water outlet is within the polygon formed by the strainers—sometimes, however, these latter are arranged as a diaphragm across the well, but, for large works, less effectively. The wire gauze supported by strong netting of large mesh, is stretched across the strainer-frames, which are best made of timber, as a galvanic action is liable to be set up between the copper gauze and iron. The frames, rectangular in shape, are made of a size convenient for being lifted and are in duplicate, balanced, sliding in vertical guides, and actuated by a rack and pinion arrangement or by winch and chain situated overhead. By this means, one set of strainers is always in action whilst the duplicate set is being cleansed.

Straining and settlement, agents of purification in almost every waterworks, were formerly the only processes adopted for the treatment of the Thames water used for the supply of London; until the increased demand necessitated the introduction of more expeditious methods, and our wider knowledge led us to insist upon more active measures being taken to purify water derived from notoriously polluted sources. Filtration, which is in the present day deemed to afford the best practical means of purifying water, whether polluted by the presence of inorganic or organic substances, is however greatly assisted and accelerated by the previous application of the processes of straining and settlement; and whilst it is cheaper to remove very fine suspended matter by filtration than by these processes, they afford in combination with filtration a more economical course of treatment than the latter alone. For, whilst the cost of constructing settling - ponds of sufficient size to permit the complete clarification of turbid water, would be frequently prohibitive; the rate at which filtration can be effectively performed is much increased, and the cost consequently reduced, by the removal of the grosser suspended matter in settling-ponds, as a preliminary step; and, as will be seen later (§ 292), the "life" of a filter, i.e. the period during which it can work without attention to the filtering medium, is the all-important factor of the cost of filtration.

274. In order to prevent the development of microorganisms of nearly every kind in water, it is desirable that water should not stand in settling-ponds exposed to atmospheric influences too long; for, whilst a goodly number of microphytes are carried down by the falling sediment, they soon begin to disperse again through clear water and to multiply under the favourable conditions that generally obtain in such ponds. There are two methods of settlement in use—the intermittent and the continuous —and the design of settling-ponds depends upon which of the two it is proposed to adopt. In the first-named system, the pond is filled and when the suspended particles are precipitated, the clear water is drawn off to a depth depending upon the length of time allotted to the action, and the degree of clarification aimed at. In order to maintain the delivery at a somewhat uniform rate, draw-off valves are advantageously situated at different levels and opened successively from the top as subsidence proceeds. By intermittent settlement, however, the time occupied in filling is partially wasted owing to the commotion caused thereby, and to a certain extent that occupied in emptying. Also, it is necessary, in this system, to situate the lowest draw-off valve of the settling-

pond at a higher level than that of the water-surface in the filters, which is sometimes inconvenient and almost always expensive.

275. Both of these objections are absent from the continuous-flow method, by which the top-water level is but little higher than that of the filters and filling proceeds simultaneously with the delivery from the outlet. One important aim of this system is to continually draw off the water that has been longest in the pond. In order to attain this end, experience shows that it is desirable to divide the settling ponds by diaphragm walls into longitudinal galleries about 6 yards wide, the depth being between 2 yards and 3 yards. A special feature of certain continuous-flow settling-ponds built at Frankfort and other places by Mr. W. H. Lindley, is a vertical diaphragm termed the "immersion-plate," sliding in grooves situated near to the outlet end, dividing the ponds into two parts. The plate is a little less in height than the depth of the water. In summer, the river-water which forms the supply is warmer than that in the pond, and, after entering, sinks as it gradually cools ; the immersion-plate is then lifted so that its upper edge stands above the surface ; consequently, the water from the neighbourhood of the bottom of the pond, which has been longest in it, is always being drawn off under the plate. In winter, on the other hand, the river-water is cooler than the pond-water, and consequently sinks to the bottom immediately it is admitted ; the immersion plate is therefore depressed, and the upper layers of water, which have been in the longest, are continuously discharged over it.

Between 15 and 20 hours are generally occupied in clarifying the water by this means, the longitudinal velocity being between $\frac{1}{14}$ inch and $\frac{1}{20}$ inch per second ; whilst at Frankfort only six hours are occupied in removing

90 per cent. of the total suspended solids. The rate of working settling-ponds depends upon the degree of turbidity of the water dealt with, and also upon the fineness and specific gravity of the sediment. The water-supply of Yokohama carries in flood time a large quantity of heavy silt that settles so rapidly that clarification takes place at the rate of 1 foot vertically in 80 minutes when perfect quiescence is attained ; and the flow through the settling pond takes place at a velocity of $\frac{1}{8}$ inch per second. Notwithstanding the extra cost incurred in construction, the advantages attached to covered settling-ponds render the type worthy of consideration. Firstly, in covered ponds the water is protected from the action of the wind, and precipitation consequently proceeds more rapidly; secondly, the water is prevented from freezing in them in winter ; and thirdly, the darkness and superior coolness are inimical to the development of *algæ* and other green plants which furnish food to animals, fungi and bacteria.

276. The construction of settling-ponds is such a simple matter that it scarcely calls for more than passing mention. The cheapest form is a sunk tank with sloping sides, the banks being formed with the spoil from the excavation of the floor. The inside of the pond may be lined with 6 inches or 9 inches of concrete, the floor being sloped longitudinally at about 1 in 30 and laterally, to a central longitudinal drain, at the same rate. The semi-fluid sludge is then readily scoured off when the tank is not working, by means of a sludge-valve outlet at the lower end of the longitudinal drain. Circumstances will determine modifications of construction which it is unnecessary to deal with here—the general features as to draw-off valves, overflows, roofing, &c., being executed much on the same lines as those of service-reservoirs (§ 262), but the tanks are of much lighter construction, because shallower.

Of course, in cases where the water is unusually clear at the source, it is most properly merely strained and filtered without settlement.

277. Filtration takes place both naturally and artificially. The greatest and also the most efficient filters are the permeable strata of the earth's crust whence our spring waters are derived (§ 131). The water that percolates into both shallow and deep wells and all underground collecting-galleries is filtered during its passage through the rocks and sub-soil, to an extent that depends upon the compactness of the strata traversed and their freedom from faults and open beds and joints. In many cases the supplies to wells or collecting-chambers gradually decrease as the pores of the superincumbent formation become choked with the mud and other matters strained out of the water in its passage.

In the case of a deep well, such decrease is usually slow ; for the water is filtered long before it arrives in the immediate vicinity of the well ; and the wide area from which the supply is drawn renders any effect from this cause insensible, and the decrease ordinarily observed is due to other circumstances (§ 129). An exception to the rule may, however, occur where water in its passage to a well, percolates beds containing very soluble substances, and, after becoming impregnated with them, partially re-deposits them in the pores of the rock around the well.

The most suitable situation for a collecting-gallery is parallel to the shore of a lake or river, on the side to which the principal subterranean drainage takes place. In such a position the gallery intercepts land-water percolating towards the lake or river, and naturally filtered through the earth. If not drawn from the collecting-gallery or well faster than it is supplied through the land, naturally-filtered sub-soil water is collected (§ 125). If, however, as is generally practised, the draught from the gallery exceeds this limit, water is drawn into it from the adjacent lake or river supposed to be at a higher level than the gallery-floor, and is, if the situation of the works is well chosen, effectively filtered in its passage through the intervening ground. When water is thus derived from rivers, if the strata constituting their beds are fairly compact, the mud filtered out is caught on the surface and is continually washed away by the current—thus a natural self-cleansing filter is established. A lake filter-gallery is without this advantage.

The quantity of water that may be derived in fairly pure condition from such collecting-galleries, can only be estimated approximately by trial borings and pumping : excessive draught means certain deterioration of the quality of the supply. The construction of filter-galleries is treated in § 125, and it is only necessary to point out the desirability of excluding surface-water by impervious sides and roof of rubble or other building-material, and of drawing the supply through the floor only.

When a portion of an aqueduct consists of tunnels, a considerable augmentation of the supply often takes place at these points, by the addition to the flow of the ground-water that drains into them. This, and the methods of collecting water by means of open canals or drain-pipes with open joints (§ 126), laid at a distance below the surface and surrounded by gravel or shells, afford other examples of natural filtration into artificial channels.

278. Notwithstanding its clearness and apparent purity, it is generally advisable that water collected from catchment-areas or pumped out of rivers should be artificially filtered before being used. Sand-filtration is the process usually adopted, although it is not infrequently supplemented by other methods of treatment. Water is filtered through sand by either allowing it to percolate downwards under gravity, or forcing it upwards through a bed of the material, which

T

varies in different works between 1 foot and 6 feet in thickness.   Such a range in the depth of the filtering layer may well call for comment, and would not be found were the action of this process of purification always clearly understood.

The system of downward filtration is that almost always practised, on account of the great facilities it affords for the cleansing of the filters—a proceeding that is greatly expedited if means are provided for forcing filtered water upwards through the sand at such a time ; this provision is most desirable, as filter-basins should always be charged in this manner.

We may here observe that in every system of purification and especially in that of sand-filtration, the maintenance and cleansing of the apparatus employed is at once the most necessary and costly part of the entire proceeding ; and true economy of design consists in devising the works so that their maintenance and the labour involved in their daily working may be as light as is consistent with efficiency.

A sand filter consists essentially of a water-tight basin containing a number of layers of the filtering material, with arrangements for admitting water above the top-layer and for drawing it off when filtered from underneath the bottom-layer.

279. There are two ways of forming the basin, the most suitable plan in any particular case depending upon the position and the nature of the foundation.   Generally, the construction of filter-basins is very similar to that of service-reservoirs (§ 262) and settling-ponds (§ 276) ; and we therefore merely recapitulate here the leading features of this class of work.   If the filter is sunk below the ground-level, provision should be carefully made for draining away land-water from the outside, at any rate down to the level of the

floor of the basin. If the ground in which it is constructed is of the nature of clay or fine sand which keeps moist by the capillary elevation of water from beneath, the basin is constructed by laying on the bottom and sides of an appropriately-formed excavation, a continuous sheet of well-tempered clay-puddle deposited and worked in the usual manner (§ 212). The floor and sides are built upon and against the puddle, of concrete, brickwork, or masonry, but need not be absolutely impervious, as indeed they seldom can be, for the puddle will prevent all leakage if properly laid. Where the basin is in cutting, the puddle behind the side-walls is supported by the solid ground ; but where it is above the ground-level, the puddle must be backed with an embankment of earth sufficiently strong to resist the pressure of water due to the head inside the basin. The side-walls must be designed of ample strength to support the lateral pressure of the puddle and earth behind them. Owing to the comparatively small head of water to the pressure of which filter-basins are exposed, and the consequent slightness of the walls and floors, there are one or two points in the construction of these latter that call for special attention. In these long thin walls the upper portion may be so far affected by temperature as to expand considerably more than the lower portion, and may exhibit a tendency to arch itself upon the latter and to form horizontal cracks along the least adherent joints of the concrete. Gaps, also, must be studiously avoided in long walls, as liable to cause the development of horizontal cracks through actual shearing of the concrete by the lateral expansion of the walls into such gaps. Similarly, during construction, the floor of a filter-basin is apt to expand into the gap formed by the recessed main drain, if that form is used, and thus to produce fracture of the drain. These difficulties, however, only arise during the construc-

tion of the works.   In hot weather, the walls may be kept
in good order by screening them from the sun ; and, as
soon as the last concrete is laid upon the floor, a few inches
of water should be run over it, the great thermal capacity
of the water effectively guarding the concrete against
excessive variation of temperature.

If the basin is completely sunk below the ground, the
sides are sometimes formed by battering the excavation to
the requisite angle of repose of the earth and wet puddle,
and covering the slopes so formed with squared stone
pitching, closely packed by hand on the top of the puddle
and not less than a foot in thickness.   This construction,
however, is better adapted to settling-ponds than to filter-
basins.   Of course, if the foundations happen to be in good
clay no puddle is required for any of the sunken portion of
the work.

280. During the long periods that sometimes elapse
when filter-basins are run off for renewal of the sand,
where the ground is very dry, puddle is liable to develop
cracks which subsequently give rise to serious leaks.   In
such situations it is desirable to avoid the use of puddle
and to build the basins of strong concrete or brickwork
or a combination of these, and to trust to the quality of
this work to secure practical impermeability.   The thickness of the floors may in such cases be made between
1 foot and $1\frac{1}{2}$ foot, and that of the walls not less than 2 feet,
but these latter must be designed of sufficient section to
withstand the thrust of earth from without (§ 203).

To prevent overflow during the process of charging the
basins with water and regulating their rate of working, an
overflow-pipe or weir may be provided at the top-water level;
but the same result is frequently attained by situating the
overflow as a branch on the inlet-pipe before it enters the
basin, or at the distributing-well from which a series of

filters are supplied, if there is one. It is convenient to fix a scour-pipe commanded by a valve at the level of the surface of the sand, for use during the cleansing of the bed, and this pipe may well be taken by a branch off the over-flow-pipe. A second scour-pipe for completely emptying the basin, drawing from the lowest point of the outlet-drain, may be provided as either a branch off the overflow-pipe or off the outlet-pipe described below, as may be most convenient. The sand employed for filtering should be heavy, of coarse rather than fine granular nature, with sharp arrises, and of $\frac{1}{30}$ inch average diameter. No sand of uniform granular size occurs in nature, and the attrition experienced by filtering-sand rounds its arrises and gradually comminutes it, until it is borne away in the washing process and lost. This waste has to be continually made good, and is a source of considerable expense in the working of sand filters. With such material, the effective straining of impurities out of the water takes place quite near the surface; and although, to a certain extent, oxidation of organic nitrites or ammonia may proceed during the passage of the water through the filter, it may be stated that generally a layer of sand 2 feet in thickness forms as effective a filtering medium as any greater thickness. The sand is carried on a thinner layer of fine gravel, which in turn rests upon one of coarser gravel, beneath which is either a bed of rubble-stone forming a collecting stratum, or a network of small drains connected with one or more channels forming the main outlet-drains, *Figs. 60* and *63*. Such channels are frequently recessed in the floors of the basins : but in order to avoid breaking the continuity of the floors the main collecting- or outlet-drains are sometimes formed as small arched culverts resting upon the floor, and pierced with holes, near the bottom, through which the water enters. If the drains as thus constructed

project into the filtering-layer, their upper part must be
built water-tight, in order to prevent partially-filtered water
from entering them, due to the rate of filtration in their
neighbourhood being increased after the filter has been
cleansed, by the diminished thickness of sand and consequent resistance.   From the ends of the main collecting-
drains farthest removed from the outlet, small pipes are
carried up inside the walls, communicating with the atmosphere immediately under the copings, for the purpose of
facilitating the escape of air from the drains during
charging of the filter.   No pipes should be carried directly
through the layer of filtering material.

The inlet to the filter is arranged to deliver the yet
unfiltered water at the level of the filtering-surface, the
sand in the immediate neighbourhood being preserved from
disturbance by stone pitching or flags set round the mouth
of the inlet-pipe ; or, sometimes, the latter enters a small
chamber built in the wall of the filter-basin, and from which
the water overflows a metal or stone sill on to the sand.
The depth of the water over this sill is kept steady by
regulating a valve on the inlet-pipe, so that the quantity of
water delivered to the filter may be maintained approximately constant.   The water entering the filter may be
conveniently and accurately gauged by a water-meter of
the inferential recording type placed on the inlet-pipe.
The Deacon meter (§ 111) is a very suitable appliance for
use under such conditions.   Sometimes, instead of the
supply being measured at the inlet, the discharge is
measured at the outlet of a filter ; but the former method
is preferable, as the measurement can be there effected
under a constant head.

281. The necessity for adjusting the inlet-valves continually, to maintain a constant delivery to filters, is
avoided when the unfiltered water is supplied direct from

reservoirs or settling-ponds, situated at an elevation sufficient to produce the required flow, and of sufficient area to prevent the surface-level from being materially affected by variation in the supply from the sources. This is a convenient arrangement in pumping-works, where the reservoir, in addition to providing the necessary balance between the irregularities of pumping which frequently occur, is constructed so as to perform effectively the work of a settling-pond. But, it must be observed, that such a reservoir by no means fulfils the complete function of a service-reservoir (§ 192), which must be provided, even in such a case, between the filters and the consumers.

The constant delivery to, and discharge from filters is necessitated by considerations of economy in the construction and maintenance of works for purification, in which it is essential that the rate of working contemplated shall be the average obtaining in the season of maximum supply ; and the equalisation of this average to the fluctuations in the demands of consumption, must evidently be effected after purification has taken place.

282. The outlet-pipe, *Fig. 60*, is taken from the end of the main collecting-drain, and is provided with a stop-valve, and frequently with a small vertical branch on the filter side of this valve, containing a float by means of which may be ascertained the "head" under which the filter is working at any time, i. e. the difference in level between the water-surfaces in the basin and the vertical pipe. As the filter becomes fouled through the interception and retention, on the surface and in the interstices of the sand, of the impurities carried by the water, the head necessary to maintain its discharge constant increases gradually, and the valve on the outlet-pipe has to be correspondingly opened, in order to ensure this increase of head by reducing the pressure in the outlet. For, it may be here

mentioned that, although not the invariable practice, it is best on every account to secure increased head by this means, and not by raising the water-level in the basin. The head is seldom, in good practice, allowed to exceed 2 feet as a maximum, depending upon the filtering medium. Before any such head as this is attained, it may be usually

Fig. 60.

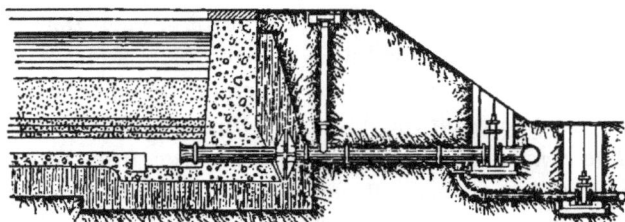

Scale 1 inch = 10 feet.

regarded as evidence that the sand needs to be cleaned, or its surface improved by the removal of the top-film.

A plan of controlling and regulating the filtering "head," superior to that of employing a throttle-valve on the outlet-pipe, is to carry the latter into a small well, which may form the converging point of the outlets of the entire series of filter-basins if at the same level. In this well, the outlet-pipes are turned up vertically, and provided with telescopic joints actuated by screws and hand-wheels overhead; and here the upper portions of the telescopic pipes are depressed, as occasion requires, to maintain constant discharge from the filters by increasing the head as the sand becomes fouled and its resistance to percolation of the water increases. This is illustrated in *Fig. 61.*

By a still more refined method, this regulation is performed automatically. Thus, at Warsaw, the discharge from the filter passes through an outlet-well, in which the clear-water supply-pipe is turned up vertically and enclosed by a cylindrical telescopic can suspended from a float, and pierced by a number of holes near to its upper edge. Whatever, then, may be the water-level in the outlet-well, the can maintains the same position relative to that level, and the clear-water discharge through the holes is

constant, notwithstanding the increased head under which the filter is being worked. At length, when the head has so far increased that the depressed water-level of the outlet-well brings the holes in the can below the up-turned edge of the clear-water supply-pipe — a condition previously arranged in accordance with the intended maximum working "head" for the filter

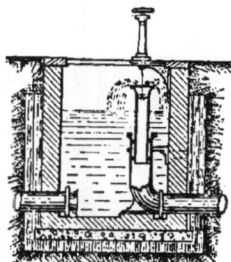

Fig. 61.

Scale 1 inch = 10 feet.

—the discharge practically ceases, indicating that the sand requires attention. A form of apparatus somewhat similar to an ordinary reducing-valve, has been also applied to this purpose.

Figs. 62 show a general plan and section of the Yokohama filters, which were designed to treat 1,250,000 gallons per day, and were constructed in 1887.

283. From the filter-outlet or outlet-well, the clear water is conducted into a "clear-water tank," which should not be less than 15 feet deep, and is preferably roofed, as light and warmth are congenial to confervoid growth ; and near large towns the presence of atmospheric impurities renders this course essential. The elevation of this tank should be such that its top-water level may be so far below that of the filter-basin, as to allow for the maximum filtering-head, together with that required for conveyance of the water from the outlet of the latter to it : and, if the distance between the two is small, the water in the clear-

water tank may be often arranged to stand, when full, at a
higher level than the sand in the filter : this is a most
serviceable arrangement in respect of charging the filter

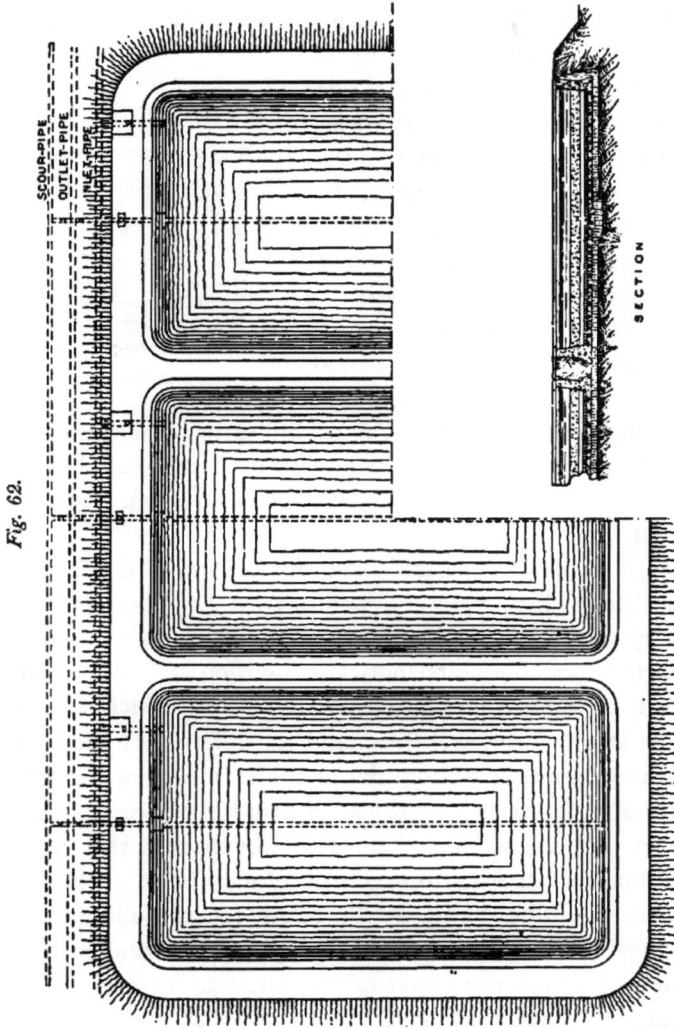

Fig. 62.

SCOUR-PIPE
OUTLET-PIPE
INLET-PIPE

SECTION

FILTER BASINS, YOKOHAMA WATERWORKS. Scale $\frac{1}{500}$.

with clear water from below, after it has been emptied for cleansing purposes, whenever it can be adopted.

Where the filters are situated near to the place of distribution, the clear-water tank may be built large enough to act as a service-reservoir ; but if otherwise, it may be quite small, built like a balancing-tank on a gravitation main (§ 357), to contain 2 hours' flow, or thereabouts, of the clear water discharged into it from the filters.

Where a series of filter-basins are situated at different levels, they cannot conveniently discharge through one outlet-well, but must each be independently provided with an outlet-well, or be throttled at the outlet-valves so as to produce constant discharge through each. The former is the better system, especially when the regulation is performed automatically ; as the manipulation of the several sluice-valves is apt to prove both expensive and inefficient in practice.

284. Having described the construction of the filter-basin and its appurtenances, we proceed to discuss the preparation of the actual filtering-bed ; because the consideration of this matter, although an essential preliminary to the design of the size and form of the basin and the arrangement of the collecting-drains, is in practice the last that claims attention, and further leads directly to the highly-important subject of the working of this system of water-purification.

In the first place, it will be readily understood that a desideratum of effective filtration, is the utmost attainable uniformity in the percolation of the water through every part of the filter. It is, however, impossible to cause the filtration throughout the whole area of a filter to be absolutely uniform, for the following reason :—

Assuming the resistance of the sand and gravel to be everywhere the same, if the percolation throughout the entire filter were perfectly uniform, the hydrostatic pressure in the drains ·underneath the filtering-material would be everywhere constant, and there could therefore be no flow through them.

In order that there may be flow through the drains, the pressure in them must decrease towards the outlet ; and, in order to make the vertical flow through the sand as uniform as possible, the pressure at successive points along a drain towards its outlet should decrease uniformly. From this it follows that, since the pressure in the drains is less as the outlet is approached, the rate of filtration is greatest near to that point, and decreases with the distance away from it.

Therefore, in order to equalise the rate of filtration at all parts of the filter as far as possible : (1) the length of the drains should be small ; (2) their area should be large ; (3) they should be numerous; (4) the size of the main drains should be increased in accordance with the number of sub-drains discharging into them; (5) the thickness of the sand should be considerable ; and (6) it should be fine in its granular structure.

The first three of these conditions would evidently reduce the fall in pressure between the remotest point of the drains and the outlet to a minimum. The fourth condition involves the fall in pressure being uniform. The fifth and sixth conditions, by increasing the resistance to the percolation of water through the sand, cause the retarding influence of the drains to be small in comparison ; consequently, notwithstanding the varying pressure in the drains at different places, the flow of water through the sand would tend to uniformity.

From these considerations it appears that, with fixed dimensions of drains, filtration is generally more uniform in a small filter than in a large one ; also, that it is more uniform the greater the thickness of the sand.

When, however, the drains are properly proportioned and arranged, the rise of pressure in them from the outlet to their remote parts is trifling in amount. The chief loss of head occurs as the water passes through the sand and the impurities which collect on its surface. But if the drains are too few and too small, it is possible that an unduly large part of the filtration may take place through that portion of the filter nearest to the outlet.

285. In filters in which the collecting-drains are formed of gravel or rubble this result undoubtedly obtains, to an extent depending on the size of the interstices of the material. The area of the drains in such cases does not increase with the increased quantity of water to be conveyed through them. Taking into consideration both cost of construction and economy of working, about 30,000 square feet is the best superficial area for filter-basins, when a large quantity of water has to be dealt with. Of course, if the total requisite filtering-area be less than this, it is necessary to proportion the size of the basin accordingly—building two filters of the required area in order that one may be cleaned and aërated when required, without interfering with the continuity of the clear-water supply. A square filter is economical in construction, because its perimeter is less than that of any other rectangular figure of equal area ; but when hollow-brick drains are employed, it is desirable that filters should be oblong—the outlet being at the middle of one of the shorter sides—as the length of the collecting-drains is thus shortened and regularity of filtration is promoted. A filter, 200 feet by 150 feet on the floor, is of economical and serviceable proportions. When, from considerations of cheapness, broken stones or gravel are adopted for the primary collecting-drains of a filter, the friction caused by them may be compensated by the provision of drains in

larger number, in order to render the flow through the sand at different parts as uniform as possible. In such a case, for a filter 200 feet by 150 feet, four collecting-drains may be provided, into which the water from the gravel primarily enters. A main drain runs up the centre of the filter to intercept the water from the collecting-drains, but that out of the gravel does not directly drain into it. In order to make the loss of head in the main drain as small as possible and so to reduce the difference of pressure at the ends of the collecting-drains, the outlet should be in the middle of the longer side. The area of the collecting-drains may be 1 square foot, and that of the main central drain such as to ensure uniform loss of pressure proceeding along it towards the outlet, i. e. in the formula (§ 66)

$$v = c \sqrt{m\,i,}$$

i would be constant.

Now,
$$\frac{Q}{A} = v;$$

therefore
$$\frac{Q}{A} \propto \sqrt{m.}$$

Assuming the drain to be square, of area A and side d,

$$A = d^2,$$

also
$$m = \frac{d}{4};$$

therefore
$$Q \propto d^{\frac{5}{2}}.$$

Denoting by d_1, d_2, d_3, d_4, the dimension of the side of the main drain after it leaves each of the four collecting-drains, respectively,

then $\quad d_1 = 2^{\frac{2}{5}};\quad d_2 = 4^{\frac{2}{5}};\quad d_3 = 6^{\frac{2}{5}};\quad$ and $d_4 = 8^{\frac{2}{5}}$

foot	foot	feet	feet
= 1·32;	= 1·74;	= 2·05;	= 2·30.

Of course, in practice, the dimensions actually employed would be the nearest to those found, that the
materials used to build the drain would work to. If in
concrete, the exact size is easily observed.

With this arrangement, the pressure in the main drain
decreases uniformly towards the outlet, and there will
therefore be a greater filtering-head near to the outlet than
away from it.

286. A superior plan is to make the primary collecting-
drains of bricks on edge covered with bricks on their sides,
or, still better, of hollow perforated bricks laid dry, through
which the water can flow with as little resistance as it
experiences in the main drains, as the following calculation
shows: Taking a filter of the same size as that previously
considered—200 feet by 150 feet, and supposing that it
filters 10 feet of water vertically per day, i. e. 3·5 cubic
feet per second; suppose the main drain, 2 feet square and
200 feet long, running along the middle of the filter,
collects all the water and conveys it to the outlet; the
velocity at the end of the drain is $\dfrac{3\cdot5}{4}$ or 0·88 foot per
second. Denoting by x the distance of any point from the
commencement of the drain, by v the velocity at that
point, by h the loss of head to that point, and by d the
side of the drain,

$$\frac{v}{0\cdot88} = \frac{x}{200},$$

or
$$v = 0\cdot0044\,x.$$

Now,
$$v = c\sqrt{d \times \frac{dh}{dx}};$$

therefore
$$\frac{dh}{dx} = \frac{v^2}{c^2 d} = \frac{(0\cdot0044)^2\,x^2}{c^2 d}.$$

Integrating,

$$\int_0^h d\,h = \int_0^{.200} \frac{(0\cdot0044)^2}{2\,c^2} \times x^2\,d\,x\;;$$

Therefore the loss of head,

$$h = \frac{(0\cdot0044)^2}{2\,c^2} \times \frac{(200)^3}{3}$$

$$= \frac{25\cdot8}{c^2}\;feet.$$

Instead of the main drain 2 feet square, suppose hollow bricks $4\frac{1}{2}$ inches square with a $2\frac{1}{2}$-inch round hole in each of them, laid end to end lengthwise along the filter-basin floor, and let the water flow through these instead of through a main drain. There would be across the filter $\frac{150 \times 12 \times 2}{9} = 400$ of these bricks; the flow through each would be $\frac{3\cdot5}{400}$ cubic feet; and the velocity at the lower end of each would be $\frac{3\cdot5}{400 \times 0\cdot034} = 0\cdot26$ feet per second.

Therefore

$$\frac{v}{0\cdot26} = \frac{x}{200},$$

or

$$v = 0\cdot0013\,x.$$

Now,

$$v = c\sqrt{0\cdot21\frac{d\,h}{d\,x}}\;;$$

therefore

$$d\,h = \frac{(0\cdot0013)^2\,x^2\,d\,x}{0\cdot21\,c^2}.$$

Therefore the loss of head,

$$h = \frac{21\cdot5}{c^2}\;feet.$$

If the hollow bricks are placed side to side with a small space between them, the area of flow is still further increased. That is to say, the loss of head which occurs in forcing the water along a central main drain is about equal to the head that would force the same quantity of water through hollow-brick drains of the kind assumed.

To summarise the requirements for causing the filtering-head all over a filter to be as constant as possible, both the primary and the main collecting-drains should be short; and no water should be led away from the point where the last collecting-drain joins the main drain by a primary drain and brought towards it by a collecting-drain; also, collecting-drains present no advantage in conveyance over perforated brick drains.

After all the water has been collected into the main drain, from that point it should be built water-tight. The head required to convey the water to the outlet from that point does not affect the uniformity of filtration.

287. The arrangement indicated in *Fig. 63* is designed to cause the least practicable difference of pressure in the drains; the water flows through primary hollow - brick drains from both ends to a central transverse col-

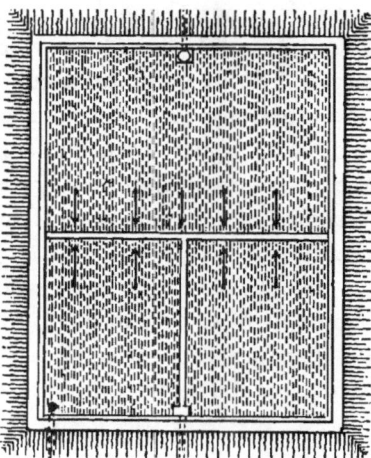

Fig. 63.

lecting-drain. The main drain being built water-tight, the loss of head in conveying the water through it does not affect the uniformity of action of the filter.

U

The loss of head in the $2\frac{1}{2}$-inch drains, from the above calculation, is $\frac{21\cdot5}{8\,c^2} = \frac{2\cdot69}{c^2}$ *feet;* and the loss of head in the collecting-drain is

$$\left(\frac{8}{6}\right)^2 \times \left(\frac{3}{8}\right)^3 \times \frac{25\cdot8}{c^2}$$

$$= \frac{3}{32} \times \frac{25\cdot8}{c^2}$$

$$= \frac{2\cdot42}{c^2} \; feet.$$

Hence the maximum difference of head with this arrangement of drains is $\frac{5\cdot11}{c^2}$ *feet,* or about $\frac{1}{25}$-inch ; with the central drain previously referred to, the difference of head is about $\frac{1}{5}$-inch. It is therefore evident that, by properly designing the drains, the loss of head from this cause is insignificant compared with the total head required to work the filter, which varies between 2 inches and 2 feet in good practice. In other words, the sand, and especially the impurities which collect on its surface and in its upper interstices, offer practically the whole of the resistance ; and it may be said that, with properly designed drains, the head is uniform over the whole area of a filter.

288. Filter-drains are made of either of the two forms shown in *Figs. 64* and *65.*

The forms shown in *Fig. 65* possess the great advantage of not breaking the continuity of the filter-basin floor.

At the end of each collecting-drain the air-pipes previously alluded to are placed to permit the free escape of air from the drains whilst the filter is being charged with water, and to preserve the sand from being disturbed by air-"blows."

289. The rate at which percolation of the water through
a filter may be permitted to take place concurrently with
efficient purification is a question of the utmost importance ;
for upon this depends the filtering-area required and the
consequent cost of the works both as regards construction
and maintenance.

Figs. 64.

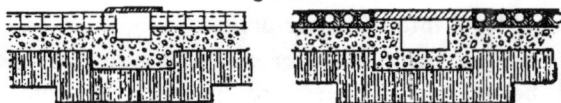

Figs. 65.

COLLECTING-DRAINS.   MAIN DRAINS.

Scale 1 inch = 10 feet.

The greater the rate of filtration, the smaller the area
required—in exact inverse ratio. It is at once clear that
this rate must much depend upon the condition of the
water to be treated ; but means are taken by precipitation
processes, accelerated, if need be, by such methods as those
alluded to in § 269, to so far clarify water that, as a matter
of actual practice, the rate may be between 8 feet and
12 feet per 24 hours.

Variations from this rule are to be found under special
circumstances ; but that expresses generally the vertical
movement of ordinary water through a sand filter. Hence
the actual filtering-area required for 1,000,000 gallons per
24 hours, lies between 13,000 square feet and 20,000 square
feet. For such rates of filtration, a thickness of 2 feet
of sand suffices. The sand employed should be angular
and free from dust ; its sharp facets will then extract the
finer particles of mud out of the water as it slowly per-

colates through the bed ; but the principal part of the straining-process is effected by the layer of fine sediment and organic scum which soon forms on the surface of the sand when the filter is put into work. The layers of gravel, usually about 6 inches thick, under-lying the sand and graduated of successively coarser quality, are introduced to prevent the finer materials of the bed from finding their way down into the filter-drains.

The normal depth of water over the surface of the sand should be merely sufficient to prevent it from being disturbed by the formation of ice in winter, and from being fouled by organic growths at other seasons—observing also that the surface of the clear-water tank may be often advantageously kept above the level of the filtering-sand. Where filters are roofed, the two former considerations cease to have weight ; but, in any case, 2 feet may be taken as the minimum depth of water over the sand in this country.

It seems ridiculous to have to state that this depth of water has not necessarily any effect upon the rate of filtration, which is entirely dependent upon the difference of the water-levels in the filter-basin and in the outlet-well, and the cleanliness of the sand and the size and arrangement of the drains ; we only do so because incorrect views on this subject sometimes find currency, and lead to improper work. The rate of filtration is, as explained in § 282, most properly governed at the outlet-well—the greatest permissible head being about 2 feet 6 inches ; beyond which, the water may be expected to cease percolating uniformly through the filter.

290. In English and Continental practice, the area of filters for large supplies varies between 2000 square yards and 8000 square yards. Economy of construction is doubtless attained by relatively large filter-basins ; but the

more potent factor of cost of maintenance—facility in cleansing, and the fact that at least one bed must be constantly out of action for this purpose, imposes a limit beyond which it ceases to be economical to increase the area of filter-basins. This limit is about 3500 square yards. It is a matter of considerable importance in designing a set of filters to arrange the basins so that the sand, as it becomes fouled, may be readily removed, washed and replaced. To reduce as far as possible carriage of the sand, the washing-machinery is sometimes placed within the basin itself: but the obvious drawback to this arrangement—that the filter has to be kept out of action whilst any portion of the sand is washed—is sufficient to prevent its general adoption. A better plan is to centralise the entire sand-washing machinery in one place, and to arrange the filter-basins symmetrically about it. The Lake Tegel filters of the Berlin waterworks, for example, are situated radially round such a space.

291. It has been already noticed (§ 284) that there is always an absence of perfect uniformity throughout a filter of the rate at which percolation through the sand proceeds, due to the diminished pressure in the drains as the outlet is approached. With a well-designed system of drains, however, this want of equality at different parts of the basin rapidly disappears as the filtering-head, in working, increases ; the loss of head in the drains should always be but a small fraction of that due to the motion of the water through the sand ; and when the latter is foul, that fraction becomes insensibly small.

The " life " of a filter is, of course, determined by the quality of the water treated, and may be found to vary between a week and two months. As soon as the head is observed to have attained the maximum prescribed limit, the basin is completely emptied of water, in order to give

the sand an opportunity of becoming aërated, and a
stratum of about one inch in thickness is removed from
the surface of the layer of sand in it.   The same method of
proceeding is employed each time that the filter requires
to be cleaned, until the layer of sand is reduced to a
thickness of 1 foot.   When this occurs, all the sand that is
discoloured, generally extending to a depth of about $\frac{1}{3}$ foot,
is removed, and the remaining thickness is then ploughed
and furrowed in order to aërate it, and so to oxidise and
destroy any small quantity of organic matter that may
have been carried down into it.   After this has been
effected, the sand is made up to its original thickness
by the addition of clean washed sand.

When the filter is ready for use, it should be filled from
below with filtered water, which should be allowed to run
away for a time off the surface of the sand.   In com-
mencing work again the rate of percolation should at first
be exceedingly slow, and should be very gradually increased
during a period of several days, until the normal rate of
working is attained, and the water first filtered should be run
to waste.   The reason for these latter precautions is that
the mud which is at first deposited on the facets of the sand,
is readily detached, until after a thin scum has been formed
on the surface, which then becomes the chief agent in
straining out and collecting the fine matter that is suspended in the water.

292. This method of starting a filter is in effect equivalent to admitting muddy water upon the sand, and then
letting the mud settle, with the outlet-valves closed; after
which the filter may be considered ready to start work.
By this means, a very thin layer of a filtering medium is
formed far finer than ordinary sand, and capable of
arresting even the minutest micro-organisms in its interstices.   If the filter-basins are at the same level as the

clear-water tank, the latter must be lowered considerably in order to empty any of the basins for cleaning purposes; and, to charge a cleaned filter from below with filtered water, the level of the clear-water tank must be high. There are thus difficulties both ways. But if the clear-water tank is also the service-reservoir, advantage may be taken of the increased consumption during the day and the decrease during the night, to effect these results.

In 10 working hours, a layer of sand $\frac{3}{4}$ inch thick can be removed from an area of about 3000 square yards by 20 men. In 4 working days of 10 hours each, the same number of men can scrape off the discoloured sand, when the thickness of the layer has become reduced to the minimum allowed, furrow the remainder, and re-fill the basin with clean washed sand. Before washing the sand that has been taken off a filter, it is well to expose it as long as convenient to the action of the atmosphere in order to allow organic matter in it to oxidise.

A method of prolonging the life of a filter, before removing the top layer of sand for washing, especially in the summer-time when a confervoid scum is apt to form on the surface of the sand, is to run off the water from the basin and let the layer of mud dry in the sun, then to break it up with rakes and wash it away by a stream of water over the surface of the sand, and down the sluice-way that is situated at that level.

293. The ratio of reserve area required to the total filtering-area depends upon the number of filters. It varies from about 1 to 7 where there are numerous filters, to about 1 to 3, in cases where there are fewer. Of course, in the case of very small filtering-areas, where only two basins are constructed, it must be 1 to 2.

As an illustration, take the case of 7 filters, each needing to be cleaned once a-week; one filter would then have to be cleaned every day, and the reserve area required would be about 14 per cent.

Upon the approach of the hottest season of the year, it is desirable to have the layer of sand made up to its full thickness, in order that, during the period in which the filters are most severely taxed, it may not be necessary to replace the sand, as this takes much longer than merely scraping and removing its upper surface.

294. As a matter of economy in working and maintenance, filter-basins situated near to a town, or in a country subject to extreme variation of temperature, should be covered.

By this means, the absorption of noxious gases and impregnation with soot is largely prevented, and a more uniform temperature is maintained in the water; although, to attain the latter result, it is necessary to have, in addition to the roof, a covering of at least $2\frac{1}{2}$ feet of earth. By the exclusion of light and the avoidance of high temperature, the development of *algæ* is prevented; consequently, the filter requires cleaning less frequently than it otherwise does. Again, in a country subject to severe winters, if the filter-basins are open, there must be provided a sufficient total filtering-area to avoid the necessity for cleansing the sand during the cold season; since, if the ice is removed from an open filter, and the basin is emptied during the prevalence of frost, the surface of the sand is at once frozen, and the whole apparatus is thereby rendered useless.

Thick ice must always be kept broken round the sides of filter-basins, as of all other such tanks, in order to prevent the serious injury to the walls that would otherwise be done by the great lateral thrust of the solid ice-sheets. The chief draw-back to covered filters is the want of continuity of the layer of sand caused by the columns that carry the roof. If these pillars are of rough brickwork or concrete, the only point to be urged against them is the difficulty of laying the sand evenly about them ; but when they are of iron, there is an additional disadvantage resulting from the tendency of the water to "creep," improperly filtered, between the sand and the smooth skin of the columns. This action may, however, be much reduced by the attachment of a collar, or diaphragm, to every column placed at about the level of the middle of the minimum layer of sand allowed in working.

The advantages of covered over open filter-basins, in point of efficiency of purification, cannot be said to have been, up to the present time, clearly demonstrated ; and are probably dependent upon the particular class of water treated. On the other hand, the advantages alluded to above are sufficient to warrant the use of covered basins in many waterworks—their cost being, roughly, 50 per cent. in excess of that of open basins of equal area.

295. Although in nearly all modern waterworks, the arrangement of sand filter-beds indicated in § 280 is followed, such is not invariably the case. Sometimes, the sand rests on but a single layer of fine gravel carried by a dry-brick floor ; the water being introduced into the bottom of the basin, is forced upwards through the filter, and flows off the surface of the sand into the clear-water tank. It is considered that, by this system, the particles strained out of the water do not penetrate so far into the sand as they do in downward-filtration systems ; it offers certain facilities for cleansing by reversed flow of the water, the inlet being closed and a scour-valve in the bottom of the basin being opened ; but its chief merit, according to our view, is that it permits of the clear water being conveniently stored over the filter, in cases where the filtering-head is derived from a main under pressure and economy of space is necessary.

296. Sand, although generally sufficient, is not always alone effective in completely removing certain very finely-divided matter from water. Turbid water that cannot be clarified by filtration through sand, may, however, be effectively treated by passing it through powdered cinders or charcoal. Thus, the water of the river Plate, which proved intractable by sand-filtration, was satisfactorily dealt with at Buenos Ayres, by passing it through a filter in which a 3-inch layer of powdered cinders was superposed upon 1 foot of sand—the rate of filtration through this medium being as high as 19 feet per 24 hours. The effect thus obtained is partly chemical, but is so largely mechanical that the filter may be properly considered here, as one in which a highly porous but more comminuted material than ordinary sand is the operative medium of clarification.

297. As has been already pointed out, the arrangements for the regular cleansing of filters, occupy a position quite as important as the construction of the works. The structural features to facilitate this operation have already been touched upon ; and we now proceed to describe briefly the processes in ordinary use for washing the fouled sand that is periodically removed from the filter-beds, as required by the exigencies of the condition of the water under treatment and its immediate influence upon the filtering-head.

A very rudimentary method, still in use to a certain extent, is to stir the sand about in a box, into which water enters at the bottom, overflowing at the top and carrying the light foul matters away with it.

Another rough method is to wash the sand down a series of shoots by streams of clean water. At the bottom of each shoot is a wire-gauze screen through which the water escapes with its burden of dirt, the sand being raked into the upper end of the next shoot, and so on successively through the whole series.

By a modification of the above methods, giving more satisfactory results, the dirty sand is placed in a box with a perforated false bottom, through which water is forced upwards and overflows at the top. The sand is stirred about by means of a horizontal revolving screw, which first turns in one direction until it has piled the sand up at one end of the box, and is then reversed, carrying the sand back again. These reversals are continued until the sand is clean. The method is open to the objection that it is not continuous in feed and delivery and is therefore costly in attendance.

An effective means of sand-washing is afforded by the revolving-drum process. The drum made of boiler-plate, 10 feet or 15 feet in length, generally cylindrical, but sometimes conical, is about 3 feet in diameter, and inclined at about 1 in 25 to the horizontal. It is fitted inside with a continuous or broken projecting spiral formed of angle-iron, making 9 to 12 complete turns in the total length. A jet of water is introduced at the upper end of the drum, or is supplied through perforations in the hollow shaft that forms its axis. The sand is supplied from a hopper at the lower end of the drum, and, as the latter rotates on its axis, is carried along towards its upper end by means of the angle-iron spiral, meeting the stream of water, which is cleaner the nearer the sand gets to the upper end of the drum. The machine is rotated at the rate of 6 or 7 revolutions per minute, delivering the washed sand continuously from its upper end. To obtain full advantage of the water, the sand is first loaded into a box, through the bottom of which the water discharged from the cylinder, enters. Thence the sand is lifted into the supply-hopper by means of a bucket- or band-elevator. By this machine, between 3 and 4 cubic yards of sand may be washed per hour; the water used being about 2500 gallons per hour.

Each drum requires about 1·5 horse-power to drive it. If the sand is very fine in quality, or extremely dirty, either a longer drum must be used, or a second may be added to work in conjunction with the first.

The number of bacteria present in foul sand may be reduced as much as 99 per cent. by such washing, if properly performed. An indifferent modification of the apparatus is that in which the drum is fixed, and the axle, armed with radial propeller-blades, revolves and carries the sand forward with it against the stream of water.

The cost of washing sand may be greatly reduced when motive-power for the washer can be obtained from the pressure in the inlet-pipes or from other hydraulic sources, instead of steam-power. It must, however, not be overlooked, that, under some special circumstances, as at Amsterdam, it may be cheaper to procure new sand than to employ washing-processes in cleansing that which has become fouled.

298. Although we have, thus far, considered only sand as the medium of filtration on the large scale, it is because we have endeavoured to deal with the subject by considering the methods of purification in order—beginning with the simpler and passing to the more complex processes.

Waters that contain not only silt, but considerable organic impurity also, in suspension or solution, are generally treated by being filtered through, or otherwise brought into close contact with, a granulated oxide of iron which has been reduced from a higher oxide; or with metallic iron that has been reduced from an oxide, and is thereby rendered in such a condition that it readily attracts oxygen to its surface or occludes it within its mass. The two well-known typical substances of this class are Spencer's " magnetic carbide " and Bischof's "spongy iron "; both of which, although in the present day already somewhat antiquated in name, are essentially the prototypes of various purifying agents now used under some other designation.

Spencer's " magnetic carbide " is magnetic oxide of iron (Fe_3O_4) in combination with carbon, and is prepared by roasting hæmatite iron ore with granulated charcoal at a dull red heat for a period of 12 to 16 hours. It is used in filters in a layer between 3 inches and 12 inches thick, according to the degree of impurity of the water under treatment; the grosser matter in suspension being filtered out by a 12-inch layer of sand above the magnetic carbide. The most important application of this medium of purification, is the well-known one at the Wakefield waterworks, where the impure water of the river Calder was purified with considerable success by its means.

Bischof's "spongy iron " is iron that has been reduced from the ore without undergoing complete fusion. It was first used on the large scale in 1880, at the Antwerp waterworks, to purify the supply derived from the river Nethe. The water was filtered through a bed consisting of a layer of spongy iron and gravel 3 feet thick, overlaid by 2 feet of sand, the normal rate of working being 22 feet vertically per 24 hours. The filtrate was subsequently aërated, filtered through sand to deprive it of ferric oxide, and was then passed into the clear-water tanks. This treatment exercised a marked effect in reducing the organic nitrogen and the ammonia present in the river-water, in addition to the suspended impurities as in an ordinary sand filter.

299. Thus far, the process seemed to be most satisfactory; experience, however, soon showed that the very

activity of the filtering medium was a serious obstacle to its continued efficiency ; for, besides the film of intercepted mud on the surface of the filters, a second deposit took place in the iron layer of the bed, consisting principally of salts of lime and magnesia and organic impurities. These deposits clogged the filters, and the difficulties resulting therefrom led to an improved method of employing iron in the purification of water being devised by Dr. William Anderson.

The apparatus, *Figs. 66*, consists of a hollow cylinder made of ⅜-inch steel plate, which is carried by hollow

Figs. 66.

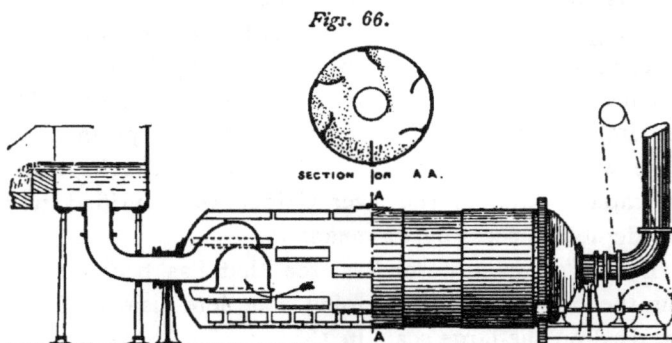

SECTION ON A A.

ANDERSON'S REVOLVING PURIFIER.
Scale 1 inch = 10 feet.

trunnions revolving in pedestal bearings at each end of the cylinder, with inlet- and outlet-pipes through which the water under treatment enters and leaves the cylinder-trunnions through stuffing-boxes. A rotary motion is given to the cylinder by means of a pinion geared into a cast-iron toothed wheel surrounding one end of the cylinder. The cylinder has five rows of short curved shelves fixed parallel to the axis of the cylinder, arranged in steps at equal intervals apart ; whilst in the place of a sixth row of shelves, is a number of square plates pivoted on spindles passing through the shell of the cylinder, and capable of being adjusted to any desired angle with the axis of the cylinder from the outside. Iron in a finely-divided state, sufficient to occupy one-tenth part of the cylinder, is introduced through a hand-hole ; and, during the revolution of the apparatus, by means of the curved shelves, the iron is carried up to the top of the cylinder, and thence is showered down through the water ; the object of the sixth row of flat shelves being to throw the iron back towards the inlet-end of the cylinder, and so to counteract the forward motion given to it by the current of water. A distributing-plate facing the inlet, prevents the water from forming a stream along the axis of the cylinder. An inverted bell-mouth on the inner end of the outlet-pipe prevents the finer particles of iron from being washed away. An air-cock at the top of the cylinder permits the escape of air during the process of charging, and also of that disengaged from the water during work. Care must be used to prevent as far as possible the ingress of air into the cylinder through the supply-pipe. The apparatus is made in various sizes—a cylinder 20 feet long and 5 feet in diameter, *Fig. 66*, requiring less than 1 horse-power to drive it, being capable of treating 1,000,000 gallons per 24 hours, allowing all the water to be in contact with the iron for $3\frac{1}{2}$ minutes. If the water is very foul, the period of contact may be increased to 5 minutes or more, the rate of flow through the cylinder being controlled by a sluice-valve on the inlet-pipe. The water leaves the purifier charged with protosalts of iron, and is aërated either by flowing over weirs or by forcing air through it ; the ferrous salts by this means become oxidised, the carbonic acid escapes, and the resulting ferric oxide is removed either by precipitation in settling-ponds, and subsequent filtration through sand at the rate of about 1 foot vertically per hour, or by the latter process alone.

The form of iron used in the cylinder depends upon local circumstances. Cast-iron borings are often employed —their loss in weight being at the rate of about $\frac{3}{8}$ grain per gallon of water treated. Cast-iron balls or small "burrs" from punching-machines, though frequently more expensive than the former, lose only about $\frac{1}{5}$ grain per gallon of water.* The particles of iron should not exceed $\frac{1}{2}$ inch in diameter in any case. Its action in the purifier is one of reduction, the subsequent oxidation of the ferrous salt being dependent upon the efficiency of the process of aëration adopted. In addition to the water being softened and clarified, the objectionable organic impurities contained in it are precipitated with the ferric oxide, and only a very small portion of them is able to pass the upper surface of the filter. This method of purification has proved generally so effective and economical, that its extended employment may be expected in many cases where the water-supply is of questionable origin.

300. Iron may be also employed in this manner to remove from water certain earthy acids, such as crenic † and apocrenic acids, produced by the oxidation of decaying vegetable matter. These are frequently present in peaty water, and in water that has percolated through a considerable depth of vegetable mould ; and, owing to their property of combining with such metals as calcium, magnesium and iron, they exercise a prejudicial influence on the animal constitution.

The effect of iron upon these acids is first to cause the formation of neutral salts soluble in water ; by exposure

* Devonshire, 'Three Years' Experience of Water-purification by means of Iron '—London, 1888.
† (Κρήνη, *a well*) so termed because they are often found in well-water.

to the atmosphere, in a second stage, insoluble basic salts
of iron are formed, and may be abstracted from the water
by the ordinary method of settlement. The process of
aëration and subsequent filtration through sand frees the
water from any excess of iron that may have been added,
in precisely the same manner as that already described
in connection with the Antwerp filters and Anderson's
revolving purifiers.

301. The desiderata of a good household filter are :—A
coarse strainer that can be easily cleansed from the
grosser impurities that it daily intercepts ; a porous but
compact mass of filtering-material, that is easily removable
but still makes a water-tight joint with the sides of the filter-
frame ; a slow rate of percolation through this material.

The simplest and most efficient filter may be thus
constructed, *Fig. 67.* The
vertical box or vessel that
contains the apparatus is
separated into three com-
partments by two internal
projecting collars. Into the
upper of these fits a frame
covered with fine wire gauze
and surrounded by an indiarubber band, so as to form a
close joint with the collar.
A block of charcoal, a few
inches thick, is set in a
similar frame and sits in the
lower collar. Both frames
are provided with rings by
means of which they may be
casily lifted out when required. The efficiency of the
apparatus depends entirely upon the care bestowed upon

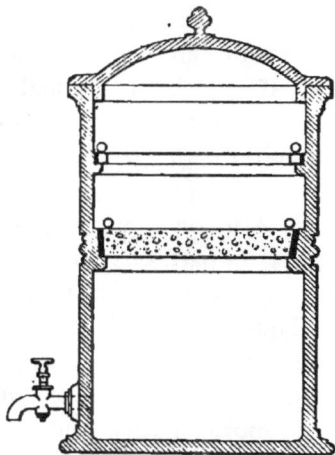

Fig. 67.

Scale 1 inch = 1 foot.

X

its periodical cleansing. The strainer should be removed and washed daily, and the charcoal slab should be re-roasted, or replaced by a new piece, at least once a-month. If, before filtration, the water is boiled for half-an-hour, the best possible result will be attained. A neglected filter becomes an incubator for microphytes, liable to damage good water if passed through it.

302. The principal effect of filtration through sand is to remove all suspended solids from the water. Nearly the whole of this matter is retained at or near the surface of the sand ; although some of the minuter particles are attracted by and adhere to the facets of the grains that lie immediately below the surface. A secondary effect, that takes place after the filter has been working for some little time, is to remove many of the micro-organisms from the water by means of the thin film or sheet of scum that collects on the surface of the sand—hence the advisability of running the first portion of the filtrate to waste if the best results are required, and of increasing the rate very gradually up to the normal, when a filter is first brought into operation. There is also frequently a slight beneficial action of the oxygen occluded in the interstices of the mass of sand, upon the organic impurities dissolved in the water.

This action is more vigorous when the filtering medium consists in part, or wholly, of carbon, which is an expensive but most valuable substance for use in small filters.

Treatment with iron, however, affords the best practicable means of abstracting the dissolved organic impurities from water ; and besides the reduction of nitrogenous matter and ammonia effected thereby, the process leads to such a destruction of the bacteria present that its use must be commended, wherever there exists suspicion that the water-supply may be a vehicle for the conveyance and propagation of pathogenic organisms.

Whilst thus alluding to the iron-treatment as a convenient practical means of sterilising water, it may be not out of place to describe briefly the nature of the evidence that is adduced when the purity of water is questioned on biological grounds.

303. The method of examination is that known as Koch's.

A certain volume, usually 1 cubic centimetre, of the water in question is operated upon thus : It is first mixed with a sterile, nutritive liquid, such as extract of meat, and sufficient gelatine is added to cause the mixture to set in a few minutes when poured over a glass plate and kept cool. The plate is then placed under a glass cover and exposed for some days to a constant temperature between 70° F. and 80° F. At the end of 4 or 5 days "colonies" of bacteria are developed of such dimensions as to be easily counted with the aid of a microscope. Further than this the determination of the species present is, to a certain extent, practicable ; and, the rapid strides now being made in this department of scientific enquiry, suggest the approach of a time when every important water-supply for dietetic purposes may be continually under inspection ; in order that the processes of purification in use may be modified as occasion requires from time to time, and adapted to deal with the various forms of impurity that claim treatment at different times.

Such water-examination, to be productive of really beneficial results, must be continuous, and form part of the routine work of the department. It cannot prudently be regarded as a mere preliminary test of the quality of the source of supply, to be applied once for all.

304. We now proceed to the consideration, in more detail, of purification by the removal of impurities in solution from water.

As a particular case, the removal of dissolved iron from the Königsberg water-supply has been already noticed (§ 269). The ground-water of the North German Plains is frequently found to be highly impregnated with iron salts, from 90 per cent. of which it is expeditiously freed by a simple process of aëration. The water is caused to fall in a fine spray upon the surface of a mass of coke broken into pieces of about 3 inches in diameter. Ferric oxide is deposited on the coke, and the water, after being passed rapidly through' a coarse sand filter, is found to be satisfactorily purified.

In this, as in some other important processes of purification, the expulsion of an excess of carbonic acid from the water, is the primary means by which the desired result is attained. Indeed, it may be said that the oxidation of nitrogenous organic matter in filters adapted to produce that result, is the only other method in extensive use, by which impurities are removed from solution in water.

The absorption, in a convenient form, of an excess of carbonic acid, forms the principle of the important process of " softening " water.

Spring-water, well-water, and that obtained from streams flowing through calcareous formations, even if comparatively free from suspended impurities, often contain dissolved mineral matter in considerable quantity, consisting of carbonates of lime, magnesia, iron and manganese, held in solution by an excess of carbonic acid, sulphates of calcium and magnesium, also alkaline carbonates, chlorides, sulphates and silicates.

The salts of lime and magnesia occur most commonly, and their presence is said to impart " hardness " to water, the amount of which property is of the gravest consequence in water used for dietetic purposes, as well as regards its economical value for industrial use. Carbonates of lime and magnesia are almost insoluble in pure water, but if it contains carbonic acid soluble bicarbonates are formed. Spring-water, and that drawn from deep wells, generally contains carbonic acid derived from the large stores of that substance which exist deep down in the earth-crust, where, owing to the great pressure, it is readily dissolved by the water.

305. In order to form a numerical estimate of "hardness" advantage is taken of the fact, that in the presence of salts of lime or magnesia, soap added to water does not form a lather, but curdles, owing to the production of insoluble lime and magnesia salts of the stearic acids in the soap. The method of measurement generally practised is that due to Dr. Thomas Clark, of Aberdeen. A standard solution of soap is made in the following manner :—Water of a known degree of hardness is prepared by first adding soap to one gallon of distilled water, until it froths on being well shaken, and then dissolving in it one grain of carbonate of lime. Instead of the gallon and grain it is in practice generally convenient to employ as measures 70 cubic centimetres and 1 milligram of the substances respectively. To render the carbonate soluble it is treated with hydrochloric acid, the solution being subsequently evaporated to dryness so as to get rid of the excess of acid ; thus the calcium carbonate is converted into neutral calcium chloride. When this has been dissolved in the water, there is present the equivalent of 1 grain of carbonate of lime per gallon of water.

A solution of pure Castile soap in dilute alcohol is then prepared, of which an ascertained measure is neutralised by the addition of a certain chosen volume, say 70 cc. of the above prepared water. The point of neutralisation is indicated by the formation of a creamy froth when the mixture is violently shaken.

The hardness of any water may be then denoted by the number of the above measures of the standard soap-solution that must be added to a volume of the water equal to that already chosen, 70 cc., in order to produce the formation of froth after the mixture has been shaken, then

Number of measures of soap − *1* = *degree of hardness.*

The unit is subtracted because one measure of soap-solution is required to make the given volume of distilled water froth after shaking.

306. When hardness is due to the presence of car-bonates of lime or magnesia, it is termed "temporary," because those salts are only kept in solution by the adventitious presence of carbonic acid. But when due to calcium or magnesium sulphates, chlorides or nitrates, hardness is called "permanent" because owing to their solubility, the latter compounds are most difficult of separation from water. In all cases, whatever particular salts may be the cause of either kind of hardness, its degree is estimated in the manner explained above; and, in this country, is expressed as so many "degrees on Clark's scale."

Temporary hardness may be removed by aëration, by boiling, or by Clark's process of adding milk of lime to the water :—

(1) By exposure to the atmosphere in cascades, or by other ways of promoting agitation, the loosely-combined carbonic acid escapes and the almost insoluble carbonate of lime is precipitated.

(2) By boiling the water precisely the same effect as that just described is produced.

(3) By adding milk of lime the additional molecule of carbonic acid is abstracted from the bicarbonate, and combines with the lime to form carbonate, thus—

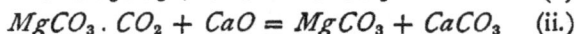

$$CaCO_3 CO_2 + CaO = 2CaCO_3 \qquad \text{(i.)}$$
$$MgCO_3 . CO_2 + CaO = MgCO_3 + CaCO_3 \qquad \text{(ii.)}$$

Permanent hardness.—The sulphates cannot be easily removed, but the inconvenience arising from their deposit in boilers may be avoided, by converting them into the extremely soluble sulphate of soda, which does not precipitate even when highly concentrated, and whose presence, therefore, does not preclude the use of water containing it for steam-generating purposes. This result is effected by adding a small quantity of soda to the water, thereby causing an exchange of acid between the calcium sulphate and the sodium carbonate ; calcium carbonate being precipitated and sodium sulphate remaining in solution, thus—

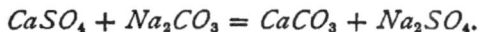

$$CaSO_4 + Na_2CO_3 = CaCO_3 + Na_2SO_4.$$

A little alum is added to expedite the precipitation of any suspended matter and of the carbonate of lime.

The quantity of lime necessary to soften a given water of known temporary hardness, is readily estimated from equation (i.) : above e. g. 70,000 gallons of water, the temporary hardness of which is 10°, would require $\dfrac{700,000 \times 14}{7000 \times 25}$ = *56* lbs. of lime added to it, to effect softening. In practice, a certain additional quantity of lime would be added, to cover loss by waste in manipulation, &c.

As an example of the removal of permanent hardness, · the water-softening works of the Taff Vale Railway Company, at Penarth Dock Station, near Cardiff,[*] treats 50,000 gallons of water per day, the reduction of total hardness being from 18° to 6°, with a consumption of 112 lbs. of lime, 25 lbs. of soda, and 5 lbs. of alum.

Water is ordinarily called " hard " when it possesses more than 5° of hardness. By the application of Clark's softening-process, the total hardness of water may often be

---

[*] Minutes of Proceedings Inst. C.E., vol. xcvii. p. 363.

reduced by as much as 15°; the residual or "permanent" hardness consisting of sulphates and other soluble salts that are exceedingly difficult of removal.

307. In Clark's process the cream of lime produced by agitating the lime with a small quantity of water in a mixer or churn, is caused to enter settling-ponds along with the hard water; here the deposition of calcium carbonate takes place, and, after the necessary interval of time, the clear water is decanted.

A valuable accessory, especially applicable to settling-ponds in which a chemically-produced precipitate is dealt with, is a decanting-pipe, which consists generally of a pipe

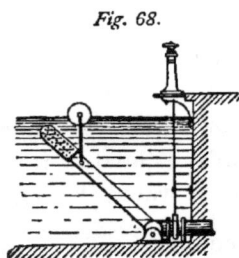

Fig. 68.

Scale 1 inch = 20 feet.

pivoted at its lower end, below the lowest draw-off level, so that its other end is capable of a considerable range of vertical motion, and furnished with a strainer and a float to keep the inlet at a definite depth below the surface of the water in the pond, *Fig. 68.* The use of this apparatus for the draw-off, ensures the clearest water being decanted during the entire period of emptying the pond.

The cost of settling-ponds is in a great measure avoided by Mr. J. H. Porter's modified process, in which the settlement takes place in a high cylindrical vessel, provided with sector-shaped shelves, arranged so as to catch the sediment as it falls, and to allow as little of it as possible to reach the bottom, *Fig. 69.* The milk of lime, and if necessary, the soda and alum solutions also, are admitted with the hard water at the bottom of the cylinder, and the chemical action proceeds and much of the precipitate settles as the water rises through it. A tank

20 feet high and 7 feet in diameter has been found capable of thus treating 30,000 gallons per 24 hours.

When the shelves require to be cleaned, a vertical shaft in the axis of the cylinder, armed with paddles fixed horizontally just above the level of the several shelves, is caused to revolve. The paddles sweep the sediment into the water which is then run to waste. When the available space is so limited that the employment of settling-ponds

Fig. 69.

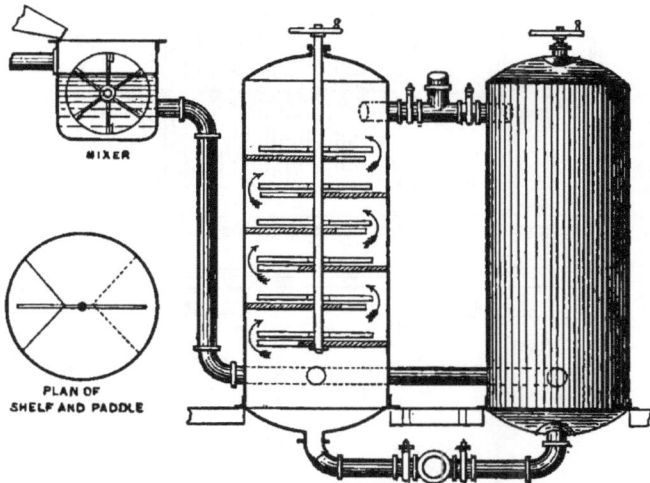

SOFTENING-PLANT—ELEVATION.
Scale $\frac{1}{80}$.

in the final stage of the process is out of the question, filter-presses, of the kind now so commonly used in sewage-disposal works, are introduced instead, to effect complete clarification of the water after it has been softened; such filter-presses are, however, much relieved if there is provided a small depositing-tank into which the water and the softening substance may enter after mixture, before passing to the filters.

The filter-press, originally invented by Mr. J. H. Porter, consists essentially of a number of parallel cast-iron plates and of rings arranged alternately in a strong frame, being carried by their projecting shoulders on a pair of horizontal rods, *Fig. 70.* Each plate is channelled circumferentially and radially on both faces, and the channels are connected with the outlet- or clear-water pipe, that is generally formed by a series of holes in the plates and rings, which

Fig. 70.

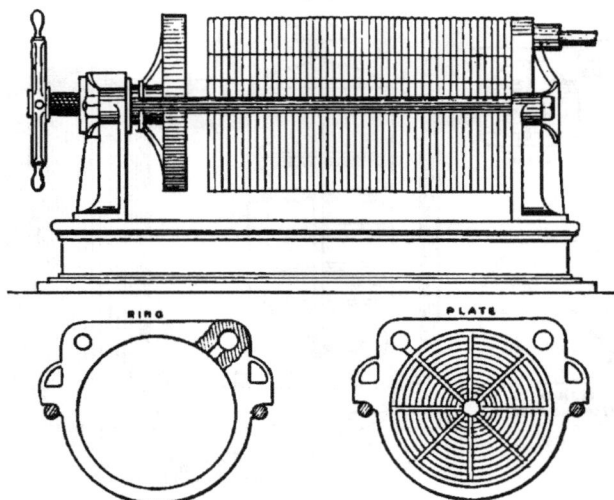

RING   PLATE

FILTER-PRESS.

Scale $\frac{1}{10}$.

can be forced into close contact by means of a screw at one end of the frame.

The feed-pipe is formed by a similar series of holes connected with each of the rings. Over each plate, and, therefore, hanging between the plates and rings, filtering-cloths of stout twilled cotton, having holes in them corresponding with the feed- and clear-water openings, are placed. The softened water is fed through the filters under a pressure of two or three atmospheres, a pneumatic pressure-pump being often used for the purpose; the carbonate of lime is retained in the ring-spaces, and the clear water passes off through the cloths and along the channelled faces of the plates, into the outlet-pipe. When the rings become filled with sediment, the screw is slackened and they are cleaned out. The cloths, especially after being in use for some time, become efficient filtering media, and as much as 30 gallons of water per hour per square foot of their active surfaces may be passed through them with good results. Many additions to and modifications of the Porter filter-press have been made for treating sewage.

308. Both large and small supplies of water are satisfactorily dealt with according to these methods, and the practice of softening is one that is daily gaining ground. Sundry specially-contrived forms of softening-apparatus and ingredients are in use, but they hardly merit description here as their principle of action is essentially that already indicated.

Beyond the immediate value attaching to water deprived by this means of an excess of the salts that produce hardness, a condition which for industrial purposes is generally desirable and often absolutely necessary, its quality as a dietetic is much improved by this, as by other processes in which a heavy but finely-divided precipitate is developed in the water. Indeed, it has been stated by Dr. Percy Frankland * that the application of these processes of softening water may effect biological purification to such an extent that the number of micro-organisms present in water may, with care in manipulation, be reduced by 98 per cent.

309. Next in importance to purifying water, is the question of maintaining the condition of purity arrived at,

* Minutes of Proceedings Inst. C.E., vol. lxxxv. p. 210.

during the interval of time that must necessarily elapse between its treatment and its supply to consumers.

It seems obvious that the shorter this interval the better; and hence we have a strong argument in favour of locating purification-works as near to centres of distribution as may be convenient; notwithstanding the economical advantage to be derived from passing only pure water through long supply-mains, which rather gives weight to the situation of such works nearer to the intake or source of the supply. It is important that naturally-filtered ground-water should be protected from the light until it is supplied to the consumers. With surface-water, this appears to be of less consequence; but it is in the highest degree essential for every kind of water intended for dietetic use, to be maintained at a low and equable temperature after it has been brought into a pure condition. We will not here anticipate the question of the storage of water in house-cisterns, and its consequent deterioration. A certain amount of unavoidable organic development takes place in the service-pipes of distribution systems, favoured by the presence of iron; and the only available safe-guard against the consequences of such pollution consists in the employment and proper maintenance of domestic filters. Experience of their management in ordinary households is, perhaps, hardly favourable to their more extended use; but, if properly constructed and systematically attended to, they may be safely advocated.

CHAPTER VI.

THE CONVEYANCE OF WATER.

310. THE construction of aqueducts has in every age formed an attractive object for the exercise of the hydraulic engineer's art. And, although the Imperial monuments reared by Caligula and Claudius in the Campagna* do not appeal to the observer's eye more forcibly than do some modern structures designed for similar purposes—our advance in the mechanical arts enables us now to frequently substitute comparatively cheap siphons of iron, hidden underground, for the imposing but costly colonnades of the ancients ; whilst, by the accurate observance of hydraulic laws, we adjust the dimensions and declivity of every portion of our aqueducts to the conditions which the physical circumstances of the locality impose.

311. The cheapest type of aqueduct is a trench or ditch cut in the earth, following the contour of the ground with a proper declivity. Such a channel, rendered water-tight, if need be, with clay-puddle, may satisfy the requirements of a conduit for the supply of water for industrial purposes —that is, where no shorter route, traversed by a more expensive aqueduct, offers advantages in point of economy. The conditions are, however, totally different in the conveyance of water for domestic use, which, once collected, must be strictly preserved from all contaminating influences. Conduits for this purpose must be impervious,

* Parker, ' The Aqueducts of Rome,'—Oxford, 1876.

and should be covered. An aqueduct pursuing the contour of the natural ground is doubtless cheaply constructed ; but its course may be so circuitous as to
deprive it of advantage over a tunnel following a shorter
line ; although the latter is generally the most expensive
kind of work for the conveyance of water. In every
case, the alignment of the route to be taken, and the
class of works most suitable to that route, can only be
satisfactorily arrived at by due consideration of an exhaustive survey of the locality, made for the purposes of
the special object in view. Precipitate selection of routes,
and hasty decision as to the types of aqueduct to be
employed, are characteristic of immature hydraulicians,
and form a fruitful source of wasteful expenditure of
money. The three principal classes of aqueduct are:—
artifical channels or conduits, pipes, and tunnels ; and
there are few works of magnitude that do not comprise a
combination of all of these types.

When the available fall is limited, it may be desirable
to employ a conduit of considerable cross-sectional area,
which, in general, may be constructed along the hill-sides
with a far less inclination than that necessary for a pipe of
equal cost and capacity for conveying water. Circuitous
routes may be avoided by crossing valleys with pipe
siphons, or by piercing hills with tunnels, according to
circumstances. The combination of these and other structures appropriate to any given aqueduct route, so as to form
an efficient and economical system, constitutes the work of
the engineer ; who must, therefore, unite a knowledge of
hydraulics to experience of that class of construction.

312. It has been already explained (§ 66) that the
flow of water under the action of gravity is determined by
the Chezy formula

$$v = c \sqrt{m\,i} \,;$$

where v is the mean velocity of the stream,

  m is the hydraulic mean depth of the stream,

   i is the sine of the surface-declivity of the stream,

and c is a coefficient.

If the surface-declivity is constant, it appears from the above expression, that the greater the hydraulic mean depth, the more favourable will be the conditions of flow. Geometry teaches that the figure which possesses the greatest area for given perimeter, is the circle ; hence, remembering that

$$Hydraulic\ mean\ depth = \frac{Area\ of\ stream}{Perimeter\ of\ channel},$$

a circle would seem to be the best form of channel to convey a stream of water entirely filling it. Under the same conditions of full flow, a semicircular channel uncovered at the top, possesses equal advantages to the circular channel of the same diameter, in point of maximum hydraulic mean depth for the amount of material required to form it.

A circular channel flowing not quite full, has, however, a delivering capacity greater than that possessed by the same channel flowing full. Using the notation of § 66 ; let A denote the area of the stream in the circular channel ;

 p „   „ perimeter of that portion of the channel bounding the stream, termed its "wetted perimeter" ;

 Q „   „ the actual delivery or quantity of water conveyed by the channel at a given time.

We have, then,

$$Q = v\,A$$

$$= A\,c\sqrt{\frac{A}{p}}\,i\,;$$

therefore

$$Q \propto \sqrt{\frac{A^3}{p}}, \quad i \text{ being assumed constant.}$$

When the channel is not flowing full, let θ be the angle subtended at the centre of the circle, whose radius is r, by the upper surface of the stream, *Fig. 71*;

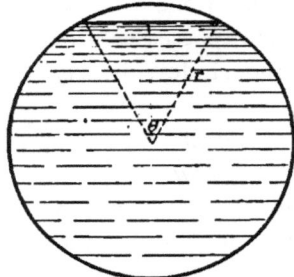

Fig. 71.

Then

$$A = \frac{r^2}{2}\overline{(2\pi - \theta + \sin\theta)},$$

and

$$p = r(2\pi - \theta);$$

substituting these values of A and p in the expression for Q, above, it is evident that Q has a maximum value when

$$\frac{(2\pi - \theta + \sin\theta)^3}{2\pi - \theta}$$

is a maximum. By the ordinary rule for maxima and minima, this is found to occur when $\theta = 51°\cdot 8.$

Hence, the maximum delivery of a circular culvert or pipe of radius r, not under pressure, occurs when the depth of the stream, in the middle, is

$$r(1 + \cos 25°\ 54')$$
$$= 1\cdot 899\ r.$$

313. Not only in channels of circular section, but in all other forms, there is a certain depth of flow at which the quantity of water conveyed is a maximum, assuming the hydraulic gradient to be parallel to the axis of the stream. Owing to practical considerations, it is generally necessary to contemplate the occurrence of full flow under pressure

in channels, at such times as their maximum conveying capacity is required to be exercised ; and, under these circumstances, the circular form of conduit is the best possible.

This condition, however, is seldom found in practice to obtain constantly. Indeed, in one class of conduits, viz. sewers, the fluctuations in the volume of flow are so great that it becomes necessary, in order to secure efficiency of working at all times, to so design the section of the channel as to give a greater hydraulic mean depth when the flow happens to be small, than that which would be afforded by a circular section of the diameter required to convey the maximum flow when running full. This suggests for consideration the form of section to be assigned to a channel in order that there may be maintained in it a constant mean velocity for varying depths of stream, the surface-declivity being constant. Such a condition must be of some consequence, if it can be realised satisfactorily in practice, in all channels that are subject to a variable flow ; but, in the case of sewers, in which the burden of solid matter held in suspension is large, it is highly important for the velocity to be fairly maintained when the flow is small and the surface-level of the stream correspondingly low.

314. With a given surface-declivity, the mean velocity of a stream varies as the square root of the hydraulic mean depth ; and it is therefore evident that, if such a form be given to the channel that its hydraulic mean depth is constant at all levels of the water-surface, the mean velocity will also be constant under all conditions of flow, assuming, of course, that the value of the co-efficient c (§ 312) does not sensibly alter ; which, under ordinary conditions, may be allowed.

Y

The fundamental equation

$$v = c \sqrt{m\,i}$$

may be written, since $m = \dfrac{A}{p}$,

$$\frac{A}{p} = \frac{v^2}{c^2 i}. \tag{i.}$$

Taking the axis of Y vertically through the middle of the cross-section of the stream, *Fig. 72*, and considering the half on the right of the figure,

Fig. 72.

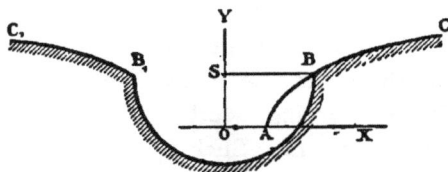

$$\frac{A}{2} = \int x\,dy,$$

and

$$\frac{p}{2} = \int \sqrt{1 + \left(\frac{dx}{dy}\right)^2}\,dy + O\,A;$$

therefore

$$\int x\,dy = \frac{v^2}{c^2 i}\left(\int \sqrt{1 + \left(\frac{dx}{dy}\right)^2}\,dy + O\,A\right).$$

Differentiating, and writing $\dfrac{1}{k}$ for $\dfrac{v^2}{c^2 i}$, (i.) above,

$$\frac{dy}{dx} = \frac{1}{\sqrt{k^2 x^2 - 1}}.$$

From this it may be seen that when $x = \dfrac{1}{k}$, $\dfrac{dy}{dx} = \infty$, and the curve is perpendicular to the axis of X at the point A.

Integrating, therefore, between the values $\frac{I}{k}$ and x,

$$\int_0^y dy = \int_{\frac{1}{k}}^x \frac{dx}{\sqrt{k^2 x^2 - I}} ;$$

therefore $\quad y = \frac{I}{k} \log_e (k x + \sqrt{k^2 x^2 - I}),$

and $\quad x = \frac{I}{2k}(e^{yk} + e^{-yk}).$ (ii.)

From equation (ii.), values of x, corresponding with all values of y, may be found; also the figure is symmetrical about the axis of Y.

For example, to find the cross-section of a brick channel of uniform inclination $\frac{I}{I500}$, such that a constant mean velocity of 3 feet per second may obtain in it, whatever be the depth of the stream : taking dimensions in feet,

let $\quad c = I00$; then $k = \dfrac{I0,000}{9 \times I500} = 0\cdot74$;

When $y = 0,\ x = \frac{I}{k} = I\cdot35$ foot

" $y = I,\ x =$ $\quad I\cdot75$ "
" $y = 2,\ x =$ $\quad 3\cdot I0$ feet
" $y = 3,\ x =$ $\quad 6\cdot30$ "

The curve A B C, *Fig. 72*, represents the profile thus calculated; and if, from the point S in the axis of Y, where the abscissa $\overline{SB} = 2\overline{OA}$, an inverted semicircle of

radius $\frac{2}{k}$ be described from B to B₁, the form of the channel
at and above B B₁ will satisfy the required condition—the
level B B₁ being, in this case, that of the lowest flow contemplated.

315. Although it possesses some features of interest
such a cross-section for a sewer could hardly be realised
in practice : nor would there be, under ordinary circumstances, any advantage in adopting it; because far higher
velocities may be permitted without risk of injury to the
walls, than can be ordinarily secured for a shallow flow in
the invert of the channel. Consequently, in practice, the
invert is formed to a small radius, so as to secure a good

Fig. 73.

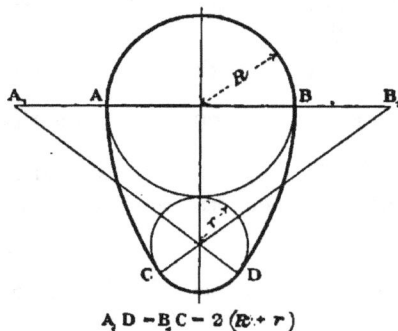

$A_1 D - B_1 C = 2(R + r)$

hydraulic mean depth relatively to the minimum flow contemplated, whilst the upper portion of the sewer is made
wider, so as to accommodate the maximum delivery.

In this way originated the ordinary oval section, *Fig. 73* ;
the general form of which possesses several practical advantages over the circular section. One of the most important
of these is the facility of access for repairs afforded by it.
It must be observed that the typical section illustrated in

Fig. 73, although almost always adopted, is in reality only rationally applicable to one particular relation between maximum and minimum delivery for any given sewer. A strict observance of the principles which led to its design originally, would require the diameters of the upper and lower inscribed circles, as well as the distance between their centres, to be determined by the local circumstances of the sewage-flow. Various economical reasons make it desirable however, to observe a similarity of section throughout any one system of sewers.

316. For ordinary aqueducts, one of the simple forms of section shown in *Figs. 74, 75* and *76*, are generally employed.

Fig. 74.	*Fig. 75.*	*Fig. 76,*
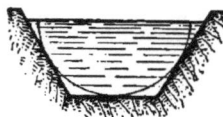	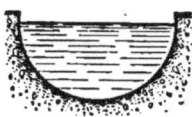	

Irrigation aqueducts are usually of trapezoidal section—the best form being one in which the sides are tangential to a circle the centre of which is situated in the water-line. If constructed of concrete, the semicircular form may be conveniently adopted.

The calculation of the dimensions and fall required for such channels to convey a given supply of water is a simple matter.

For example, consider a channel of the semi-hexagonal form illustrated in *Fig. 74*. If r is the radius of the inscribed semicircle, the area of the semi-hexagon is $\sqrt{3}\,r^2$, the wetted perimeter of the full channel is $2\sqrt{3}\,r$, and the hydraulic mean depth is therefore $\frac{r}{2}$—the same as for the semicircle.

Let v denote the mean velocity of the stream, which, for masonry or brickwork channels, may be 4 feet or 5 feet per second, but for concrete should not exceed 3 feet and for earth 1·5 foot per second. Let Q denote the delivery of the stream in cubic feet per second, and let c be the coefficient proper to the case (§ 67).

The required sectional area of the stream is

$$\frac{Q}{v} = \sqrt{3}\, r^2 \, ;$$

from which r is found.

To find the surface-declivity ;

since
$$v = c \sqrt{m\,i}, \text{ and } m = \frac{r}{2},$$

$$i = \frac{2\,v^2}{c^2 r}\, ;$$

That is
$$i = \frac{2\,v^2}{c^2 r} \times 5280 \text{ feet per mile.}$$

317. It frequently happens that the ground which has to be traversed by the conduit limits the surface declivity of the stream, and we require the dimensions of the channel, which, with that declivity, will convey the desired supply of water. Since the hydraulic mean depth of the channel, which is a factor of the value of c appropriate to this case in the fundamental equation, is unknown, the question cannot be solved directly. The following example will show the method of treating this case, more clearly than a general explanation of it.

Example.—Required the dimensions of a rubble masonry

channel to convey 2 cubic feet of water per second, the maximum available declivity being $\frac{1}{2500}$, and the geometrical form of the cross-section being a semi-hexagon.

Let the desired mean velocity in the channel be 1 foot per second, whence A is found to be 2 feet. Assume a value of c,—say, 80.

By (i.), § 314,
$$\frac{A}{p} = \frac{2500}{6400} ;$$

substituting the above value of A,

$$p = \frac{128}{25} = 2\sqrt{3}\, r ;$$

therefore
$$r = 1\cdot 48 \text{ foot,}$$

and
$$m = 0\cdot 74.$$

Introducing this value of m into the equation § 67, we obtain, see Table of values of c,

$$c = 81.$$

Hence, as a second approximation,

$$p = \frac{131}{25} = 2\sqrt{3}\, r.$$

Therefore the radius of the inscribed circle of the required channel is, for practical purposes, $1\cdot 5$ foot.

318. The foregoing considerations are in general equally applicable to all kinds of aqueducts—whether open channels, pipes, tunnels or culverts. Every aqueduct must be designed to suit the hydraulic gradient that obtains naturally between its extremities. This gradient, in all ordinary practice, comprises two principal elements: (1) the head of water employed initially at the commencement of the aqueduct, to produce the required mean velocity of flow in it; and

(2) the head employed in overcoming the frictional resistance opposed by the surface of the channel in contact with
the stream to the motion of the latter, throughout its entire
length.

In the study of the pure science of hydraulics, a third
element is considered: viz. the head lost at the entrance of
the conduit by those internal motions of the water which
result in the formation of *vena contracta.* In good practice
such loss is rendered insensible by the employment of
curved or trumpet-mouthed inlets, and need not be considered here.

Both of the former elements are important, especially
where a considerable portion of the total head is applied to give motion to the
water, at points of the aqueduct at which its flow has
been checked or temporarily
stopped by the interference
of wells or weirs.

Fig. 77.

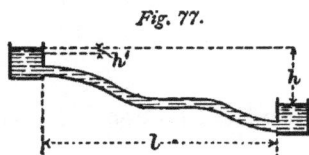

If h denotes the total fall in an aqueduct of length l,
h' the head required to produce the initial velocity v, and
h'' that necessary to overcome the frictional retardation of
the flow, throughout its entire length, *Fig. 77*: The declivity or hydraulic inclination is

$$\frac{h - h'}{l},$$

and since

$$v = c\sqrt{m\,i},$$

therefore

$$v^2 = c^2\, m \frac{h - h'}{l}; \qquad \text{(i.)}$$

also

$$h' = \frac{v^2}{2\,g}.$$

Let n be the number of times that the flow is started from a condition of rest, in the length l considered; then $\dfrac{n\,v^2}{2\,g}$ may be written for h' in equation (i.);

therefore
$$v = c\,\sqrt{\dfrac{m\,h}{l + \dfrac{n\,c^2\,m}{2\,g}}}\,.\qquad\text{(ii.)}$$

This expression (ii.), into which the value of c appropriate to the circumstances, (§ 68) is introduced, is generally applicable to all kinds of aqueduct, of which the length, total available fall, and form of the cross-section are known.

319. In addition to the simple ditch cut in the earth, puddle-lined, or it may be hand-pitched with stone, as required by circumstances, and the various more or less elaborate structures of masonry, brickwork or concrete, conduits are, in some special cases, constructed of iron, e. g. the Pont-Cysylltau near Ruabon, built by Telford in 1803, or of timber, as is a common practice in the United States at the present day. Some of the timber "flumes," or troughs, used for irrigation purposes in that country, are of extraordinary magnitude, one of the high-level trestle-work conduits of the Platte Canal measuring 1000 feet in length, and 30 feet by 7 feet in cross-section. These flumes are frequently carried for great distances upon shelves or benchings cut in the hill-sides—a form of construction which is found, where timber is plentiful, to be cheaper than stone- or concrete-lined ditches. It is desirable for a long aqueduct to have sluice-doors, with waste-weirs on their up-stream sides, at intervals throughout its length, so that, in the event of accident, or need arising to examine any section of the work, the water may be cut off above that portion and run to waste over a weir

into some convenient water-course. A simple iron or wcoden shutter, sliding in vertical grooves, and actuated by a rack and pinion, forms an effective sluice for ordinary conduits. Large sluices for use in rivers and canals require a very different kind of construction, but are hardly within the scope of this work.

320. Although aqueducts such as those alluded to in the preceding sections, frequently afford favourable means for the conveyance of water, as witness the conduits on the line of the Glasgow water-supply from Loch Katrine, and that contouring the hill-sides of Grasmere and Rydal on the route of the Thirlmere water-supply to Manchester, now under construction; by far the most generally serviceable means of conveyance, if circumstances permit its adoption, is the iron pipe. Being competent to withstand high internal pressure, a strong iron pipe may pursue a direct line across the country, following the natural undulation of the ground, provided that it nowhere rises above the hydraulic gradient. In this country, water-mains are usually made of cast-iron; but, abroad, the exigencies of bad roads and indifferent means of locomotion frequently cause the introduction of wrought-iron or steel tubes. Pipes of the last-mentioned materials, although of less weight, are more expensive in substance and workmanship than those made of cast-iron, and, usually, but little gain in point of cost can result from their employment. They cannot be laid in positions where they may be subject to the passage of heavy weights over them, and they must, if carried above ground, be shielded from sun and frost by timber or earthen covering. On the other hand, the rigidity of cast-iron pipes allows of their being laid underground, where atmospheric influences cannot affect them, and, where opportunities offer, cast-iron mains may be thus carried along the public highways. This is

sometimes an important advantage, because by it, many questions of damage and severance, incidental to the construction of a pipe-line through private property, are avoided.

321. Where large supplies of water have to be conveyed the question arises whether one large main, or two or more of smaller size, should be used. If the question were one of prime cost only, it would generally be decided in favour of the former plan. With a given inclination, the delivery of pipes varies approximately as the fifth power of the square root of the diameters, whereas the weight and consequently the cost of the main varies as the square of the diameter. Hence, the delivery, which varies simply as the total cost, when a number of similar mains are employed, bears a higher relation to the cost when a single large main, of conveying-capacity equal to the aggregate of these latter, is used. Other questions, such as the damage resulting from the bursting of a large main, the size and strength of the valves required, and the inconvenience arising from temporary cessation of the supply during repairs to any part of the pipe or apparatus, exercise in an important degree the judgment of an engineer in deciding this matter. Cast-iron mains of 50, and even 60 inches diameter have been successfully manufactured and used under pressure ; but present practice does not favour the employment of mains much exceeding 40 inches in diameter for ordinary purposes.

In deciding what size of pipe to adopt, the important question generally arises—what capacity should the pipe have, in order to meet future increased demands for water to a suitable extent, without laying an inequitable burden of cost upon the present undertakers of the works ? It is evident that the answer to this question must be largely given by individual opinion ; but it is clear that the

accumulated interest of any capital outlay upon such parts
of the works as are only partially used during a long period
of time, should not exceed the cost of equivalent additional
new works when the need for their actual employment
arises. In great water-supplies it is customary to lay a
series of mains side by side, at such times as the increasing
demand for water requires it; but for small supplies, the
propriety of this course must depend to a considerable
extent upon the length of the main in question and other
practical and fiscal conditions.

322. When the daily rate of supply to be provided for
is known, and the route of the pipe-line has been selected,
the design of a main is a simple matter. The velocity
permissible in the largest class of pipes is about 3 feet per
second; though, for very short lengths of pipe, this may be
considerably exceeded. In service-pipes of 2 inches or 3
inches, a mean velocity of 10 feet per second may be allowed
if proper precautions are taken to prevent sudden arrest of
the motion. Between these limits, the velocity may be
safely taken varying inversely as the square root of the
diameter of the pipe.

To find the diameter or the declivity of a main when
either of them is given: Let v denote the mean velocity in
the pipe; divide 4 times the supply, reduced to cubic feet
per second, by πv,—then the square root of this quantity
gives the diameter of the pipe in feet.

Thus, if Q denotes the supply in cubic feet per second,

and D „ the diameter of the pipe in feet,

$$Q = \pi \frac{D^2}{4} v ;$$

therefore
$$D = \sqrt{\frac{4\,Q}{\pi\,v}}.$$

As the hydraulic mean depth of a pipe flowing full is $\dfrac{D}{4}$, we obtain, by transposition of the fundamental equation $v = c\sqrt{m\,i}$ (the values of c being those given in § 72) :—

$$D = \frac{v^2}{3000\,i} \text{ for new iron pipes,} \tag{i.}$$

and

$$D = \frac{v^2}{1500\,i} \text{ for rusted iron pipes ;} \tag{ii.}$$

or

$$i = \frac{v^2}{3000\,D} \text{ for new iron pipes ;} \tag{i. } a$$

and

$$i = \frac{v^2}{1500\,D} \text{ for rusted iron pipes.} \tag{ii. } a$$

From these equations, the required diameter of the pipe, or the hydraulic inclination, as the case may be, can be found. In this calculation, it is necessary to observe that engineers have generally to contemplate the conveying-capacity of mains when it has become reduced by fouling and incrustation, to which, notwithstanding all precautions, every iron pipe is in the course of time liable.

It is desirable therefore to use the expressions (ii.) and (ii. a), and, further, to increase the diameters assigned by calculation to pipes, by an allowance for incrustation. The amount of this allowance depends largely upon the quality of the water to be conveyed, and is, of course, equal to the incrustation that it is in contemplation to permit, before scraping or otherwise cleaning the mains. This precaution, too often neglected—especially in small pipes, which are naturally subject to a vastly more rapid relative diminution of area from this cause, than are those of larger diameter—will prevent much of the disappointment that is sometimes experienced in small waterworks designed by unskilful persons.

It should be almost unnecessary to point out, that the probable future demand for water must receive the most earnest attention, when the diameter and declivity of a supply-main are under consideration ; for the carrying-capacity of a water-pipe, unlike that of a railway or a road, cannot, when once laid to meet certain definite requirements, ever be safely augmented by increasing the velocity of transit, even where facilities for effecting this occur. The matter is one upon which no sound judgment can be formed without experience.

323. The thickness of a water-pipe, regarded as a thin cylindrical shell, to resist internal pressure, is arrived at thus :

Let p denote the intensity of the internal pressure to which it is exposed,

s „   „   working load of the metal,

t „   „   required thickness ;

then
$$t = \frac{p\,D}{2\,s},$$

where D is the diameter of the pipe ; or, denoting by w the weight of unit-volume of water, and by h the total head on the pipe in question,

$$t = \frac{h\,w\,D}{2\,s}.$$

If, as is customary, t is measured in inches, D and h in feet, w in tons, and s in tons per square inch ;—

$$t = \frac{h\,D}{2 \times 12 \times 36 \times s}$$

$$= \frac{h\,D}{864\,s}. \tag{i.}$$

But if the value of t thus found by (i.) for any pipe, is

less than a certain amount, shown by experience to be necessary for the production of a sound casting and the preservation of rigidity under the usage to which the pipe is subjected in work—especially the external pressure brought to bear upon it by the earth when it is laid—its thickness is increased up to that amount, which may be determined by the empirical rule

$$t = \frac{\sqrt{D}}{2}.$$

Example.—If a cast-iron pipe, 3 feet in diameter, under a head of 250 feet of water, is 1 inch in thickness, the additional thickness t' of metal required to enable it to withstand a head of 300 feet, would be—assuming a working strength of 1 ton per square inch in the iron—

$$t' = \frac{50 \times 3}{864}$$

$$= 0\cdot 17 \; inch \; ;$$

that is, the pipe would have to be made 1·17 inch thick.

Owing to the practical impossibility of obtaining perfectly cylindrical and concentric pipe-barrels, it is proper to add to the thickness found above an amount varying from $\frac{1}{16}$ inch for small pipes to $\frac{1}{8}$ inch for those of larger size—this being the maximum deviation from the specified thickness to be permitted in manufacture.

Every portion of a main under pressure must be calculated with respect to the maximum pressure that it will have to sustain, in order to attain an economical result. The pipes comprised in a long main are grouped in classes, and generally vary by $\frac{1}{8}$ inch to $\frac{1}{16}$ inch in thickness—by the results of the calculations.

The external diameter in each class is constant throughout, and the "steps" in the thickness of the barrel are effected by varying its internal diameter. By this plan, equality of the joints is secured, with a uniform size of socket ; and the number of " taper-pipe " specials is reduced. If the main possesses a wide range in the thickness of the pipe-barrels, a taper-pipe should be introduced at every fourth step, in order to avoid too great variation of the internal diameter. In connection with this part of the subject, it may be remarked that no little attention is required in the practical operation of laying the main, to ensure the proper position of each class of pipes being duly observed.

324. In the supply of high-pressure water for hydraulic power, the forces dealt with are so considerable—frequently pressures exceeding 700 lbs. per square inch—that the thickness of the pipes employed bears a high ratio to the diameter, and the barrel cannot be treated as a thin shell in computing its strength.

Analysis of the conditions obtaining in a thick hollow cylinder subject to internal fluid-pressure leads to the equation

$$\frac{p_0}{q_0} = \frac{R^2 + r^2}{R^2 - r^2} ;^*$$

where R, r, are the outer and inner radii, respectively,

p_0 is the maximum hoop-tension, say, one-eighth of the ultimate tenacity of the metal,

q_0 is the radial pressure on the inner surface of the cylinder ;

therefore
$$R = r \sqrt{\frac{p_0 + q_0}{p_0 - q_0}} .$$ (i.)

From (i.) r, p_0 and q_0 being known, R is found ; and from this $(R - r)$, the thickness of the barrel of the pipe, is determined.

* Rankine, 'Applied Mechanics,' 9th edition, Art. 273.

Thus, $$t = r\left(\sqrt{\frac{p_0 + q_0}{p_0 - q_0}} - I\right).$$ (ii.)

Example.—Taking dimensions in inches, and stresses in lbs. per square inch, a high-pressure main, 6 inches in diameter, worked at 600 lbs. per square inch, must have—assuming as the maximum working stress in the metal, 2200 lbs. per square inch—a thickness of

$$3\left(\sqrt{\frac{2200 + 600}{2200 - 600}} - I\right) = 0\cdot 96 \; inch :$$

Adding the amount necessary to cover defects of figure in casting, the pipe-barrel should be 1 inch thick.

325. In the conveyance of water for power purposes, it is important, at the outset, to know how far the mechanical efficiency of the apparatus to be employed for the transmission of power, whether electric, pneumatic or of any other kind, is greater or less than that of the aqueduct in which the water is to be conveyed. In other words, will less head be absorbed in conveying the water, or in transmitting the corresponding power in the form of electric, pneumatic or other energy? The answer to this question must influence in an important degree, the situation of the works where the transformation from water-power to the desired form of energy, is effected—whether near to the source of supply, or at the place where the power is to be utilised, if these are some distance apart.

With regard to the size of pipe to be used (we do not here speak of the transmission of hydraulic power from a generating-station throughout a district of supply) to convey water from a natural source of supply to a place where its potential energy is to be converted into, say, electric current, the question arises, is it cheaper to employ a large pipe and a comparatively low pressure or a smaller high-pressure main conveying equal energy?

Approximately (§ 323) the weight of a pipe, and therefore its cost, varies as its diameter and the pressure to
which it is subjected ; then, for a given expenditure upon
a pipe, the pressure under which it may be worked varies
inversely as its diameter.

Its capacity for conveying water with a given loss of
head varies as its (diameter)$^{\frac{5}{2}}$.

The water-power that can be conveyed by it is the
product of its conveying-capacity and the pressure, and
therefore varies as

$$\frac{diameter^{\frac{5}{2}}}{diameter} = (diameter)^{\frac{3}{2}}.$$

From this it appears that the water-power that can
be conveyed by a main varies at a higher rate than its
diameter, *ceteris paribus* ; and, therefore, where the efficiency of the water-motors is no greater at high than at
low pressures, it is economical to convey water at low
pressure with a correspondingly large diameter of pipe—
assuming that the quantity of water available is such as to
permit this disposition of the matter, within the limits
imposed by the mechanical conditions that govern the
efficiency of water-motors.

The case is, however, entirely different for the distribution of hydraulic power from a pumping-station throughout a neighbouring district. Here the cost of the development of the power is so considerable in relation to that
of its conveyance, as to affect in an important degree
the question of the most suitable pressure for the water-
mains. The experience of the great hydraulic-power
installations of London and Liverpool is sufficient to prove
that the high-pressure system is absolutely the best under
such circumstances.

326. The joints of cast-iron pipes are formed in several
ways :

Figs. 78 and *79* show ordinary forms of "lead" joint—
the annular space between the spigot of one pipe and the
socket or "faucet" of its fellow, into which it is introduced
to a depth of about 4 inches, containing a ring of lead
tightly caulked in round the face. The lead-space is
slightly coned towards the face of the joint. "Turned-
and-bored" joints are illustrated in *Figs. 80* and *81*—the
joint being formed by the adhesion of the turned surface
of the spigot of one pipe against the bored interior of the
socket of the next one. The prototype of this joint is

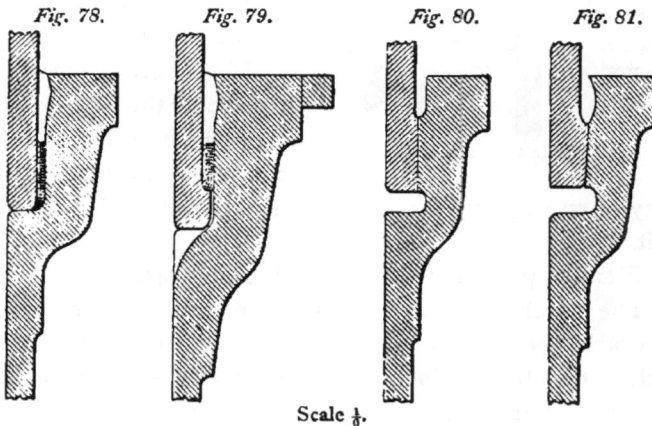

| *Fig. 78.* | *Fig. 79.* | *Fig. 80.* | *Fig. 81.* |

Scale ⅛.

to be found in the ancient wooden water-pipes made of
tree-trunks pointed at one end to a spigot-shape and bored
out—the bore being enlarged to form a faucet at the other
end. Large numbers of such pipes are still used in districts
where economy recommends them before other considerations. For example, in Detroit, Mich., 87 miles of these
timber pipes might have been found a few years ago, some
of them laid so recently as 1880.*

Figs. 82 and *83* give "flange"-joints; the actual joint
being formed by the close meeting of the planed fillets

* Report on the Water-Power of the U.S.—Washington, 1887.

Z 2

1½ inch wide that project about about ¼ inch beyond the flanges, and are drawn together by the tension of the bolts. The fillets are sometimes omitted, and the faces of the flanges are planed right across.

The joint, *Fig. 84*, has proved effective in high-pressure water-mains.   It is of a form introduced a few years ago

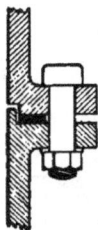 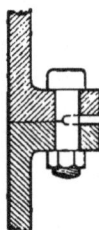 

Fig. 82.   Fig. 83.   Fig. 84.

Scale 1 inch = 1 foot.

by Mr. E. B. Ellington, which has successfully overcome the difficulties attending the application of great pressure in drawing the flanges together.

The actual joint is formed by the indiarubber ring which is forced into the recess on the extremity of one pipe by the fillet on the extremity of that next to it.

Flexible pipe-joints are alluded to in § 345.

The depth of "lead" joints depends upon the pressure to which they are to be subjected; but the lead ring is seldom less than 2 inches, or more than 2½ inches broad, and has a thickness of ¼ inch. The thickness may, however, range up to ⅝ inch in wide-socket pipes, employed sometimes for laying mains round curves. The total depth of the socket is about as much again as that of the lead in the joint; the space behind which is filled with either spun-yarn or lead-wire, jammed hard against the back of the socket before the molten lead is poured into the joint.

Large pipes are sometimes furnished with half a dozen equidistant lugs cast on the back of the sockets, inside, to provide a bearing for the spigots of those following them ; but these lugs have generally to be hand-dressed, or cut off altogether when the pipes are working round a curve.

Still, they are useful in preventing, to a certain extent, the sockets from being burst by the application of too much force in driving the spigots home. The foregoing remarks apply both to ordinary straight pipes and to the numerous special pipes: Bends, collars, branches, taper-pipes, double-socket, double-spigot, flange-and-socket, and all pipes necessary for the introduction of valves and other appurtenances of water-mains.

Straight pipes are usually, if less than 1 foot in internal diameter, cast in 9-foot lengths ; above that size they are cast in 12-foot lengths. Special pipes may, with advantage, be made one-half of these lengths.

327. In the best English practice, pipes are made of mine-pig iron, re-melted in a cupola, and are cast vertically, socket downwards, if they are of spigot-and-socket type, with a head of 1 foot of metal which is afterwards cut off in the lathe. In planing the flanges of those pipes that possess them, care must be taken that the faces are truly at right-angles to the axis of the pipe. The usual requirements for the production of sound castings are to be specified in contracts for manufacturing them. Their cylindrical shape is proved by rolling them on an iron gauntree, and their straightness and concentric figure by callipers and straight-edges. The spigots and sockets are tested as to dimensions and form by circular steel template-rings.

328. The tests applied to metal used for the manufacture of cast-iron pipes are various, and are not always consistent with one another: one proof of strength which has been applied to many thousands of tons of iron for pipes within our experience, with satisfactory results, is that a bar of metal, 40 inches long, 2 inches deep, and 1 inch wide, the weight of which must not exceed 21 lbs., shall, when supported on edge at points 36 inches apart, sustain a load of 3000 lbs. applied at the middle of its

bearing for 1 hour, and shall under this load deflect at least
$\frac{3}{8}$ inch at the middle.   An additional test of tensile strength,
which has much to recommend it, is that a bar of the metal,
8 inches long and 1 inch square in section, shall sustain a
load of 8 tons (tensile stress) for 1 hour.   As an alternative,
the transverse stress may be slightly reduced and the
longitudinal test be somewhat increased.   After they have
been examined, the pipes are proved with oil in a testing-
machine, *Fig. 85.*

The ordinary form of this apparatus consists of two
vertical iron shields, between which the pipe is placed
horizontally and in one of which is an inlet for the oil
furnished with a pressure gauge.   One of the shields is

Fig. 85.

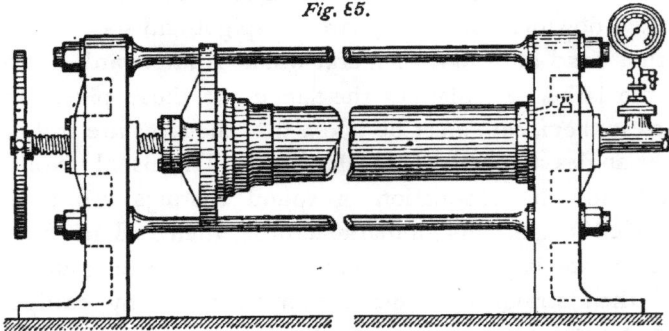

fixed and the other is movable, the latter being forced
against the pipe by either hydraulic- or screw-pressure—the
joints being made with leather or indiarubber rings.   The
pipe is then fully charged with oil, and the proof-pressure,
which is generally double the working-pressure, is applied
by a force-pump and is maintained for 2 or 3 minutes—
the barrel of the pipe being meanwhile repeatedly struck
with a light hammer and closely examined for any trace of
leakage.   If proved sound, the pipe is at once re-heated in
a stove to a temperature of 600° F., and is then dipped

into a solution of pitch, asphalt, resin and linseed oil, at
a temperature of 300° F., in accordance with the process
invented by the late Dr. Angus Smith of Aberdeen. The
film of varnish thus laid on, rapidly hardens, and is, if
applied to clean metal, possessed of remarkably adhesive
and durable properties—preserving the iron from corrosion
for many years. But if, as too frequently happens, the
castings are exposed to the atmosphere before being coated,
and so become oxidised, the whole process is useless and
might as well be dispensed with. Coating pipes with gas-
tar or other similar substances is also a futile operation.

329. With regard to wrought-iron or steel pipes, the
smaller sizes are drawn or welded, and
those of larger diameter are riveted.
The joints are flanged and bolted,
jammed spigot - and - socket - wise, or
collared with single or double lead-
joints, *Figs.* 86, 87.

Fig. 86.   *Fig. 87.*

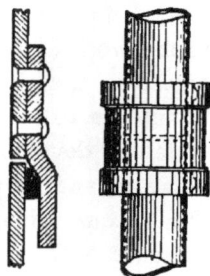

Scale 1 inch = 1 foot.

The joints illustrated in these figures
possess the advantage of allowing for
the natural expansion and contraction
of the main ; and it may be observed
that, where flanged, or turned and
bored, or any other rigid joints are
used for a cast-iron main, it is desirable to make, say, every
tenth joint of lead on this account.

A preservative coating, consisting of Stockholm tar and
red lead, may be advantageously applied to wrought-iron
and steel pipes, which must, however, be scraped perfectly
clean before being thus varnished. The mixture is prepared by adding red lead to the heated tar, stirring it well
until it contains as much as is consistent with the fluidity
necessary for its easy application with a paint-brush. A
wash of Portland cement forms a cheaper coating, and is

found to answer fairly well.   After the pipes have been coated and dried, they are weighed—their weights affording an important check upon the section of the metal in the barrels.   It is customary to specify a minimum weight for every class of pipe, in addition to the specification of the normal thickness and permitted deviation therefrom (§ 323); also a maximum weight, beyond which any excess of metal is not paid for.

330. The most minute attention must be paid to the manufacture of water-pipes, during every stage of their progress from the foundry to the trench in which they are finally laid.   Without regular and intelligent ¦inspection, good results are unattainable.   After the pipes have left the manufacturer's hands, they are often exposed to severe vibration and to rough usage, in the course of their transit to the works.   Blows or pressure are apt to produce fine cracks in the skin of the metal, which are frequently so minute as to escape observation ; but which will be to a certainty developed by vibration or by variation of temperature, and may lead to fracture of the pipes in which they occur, when the latter are exposed to the water-pressure.   Striking the castings with a hammer, when slung in the chains over the trench, will not indicate the presence of cracks that only extend slightly below the surface of the metal.   We have repeatedly observed such small cracks extend right through strong hooped sockets by the expansion due to the mere warmth arising from the process of "running" the lead joint.

As a precaution against fracture of the spigots, which are the weakest portion of pipes, during transit, wrought-iron rings are often shrunk on them ; and these have in our experience proved effective—especially in cases of the carriage of pipes by sea, where, from the rolling of vessels, they are exposed to greater risks of damage than in perhaps any other kind of transit. In large pipes, wrought-iron hoops are sometimes shrunk on the sockets, to add strength to enable them to withstand the somewhat severe tensile stress caused in them by the caulking of the joint.

Nothing, however, short of expert inspection at the trench-side, and rigid adherence to the rule of not passing a single flaw, can be relied upon to ensure a sound main. It is not too much to say that, were this practice invariably adhered to, the " prevention of waste " would play a much less important part in the maintenance of waterworks than is the case at present.

When pipes are fractured at the ends, they may be cut to shorter lengths, and so rendered serviceable. They are best cut by forming a chase round them with a diamond-pointed chisel, and then driving, simultaneously, sharp steel wedges through the barrel at three equidistant points in its circumference. By this method, the pipe is separated into two parts with a clean cut following the line of the chase.

In some cases, single small air-holes may be drilled out, and if the barrel appears to be otherwise sound, a thread may be formed in the hole, when a brass or gun-metal plug is inserted. Such patching as this is only permissible with the express sanction of the engineer, and under his inspection ; and, in any case, the pipes so treated should be subsequently proved under pressure before use.

Cast - iron pipes are manufactured of sizes varying between the maximum limit of, say, 50 inches in diameter down to a minimum of about $1\frac{1}{2}$ inch. Owing, however, to the difficulties of varnishing small pipes internally, and the rapid diminution of their area with but little corrosion, such pipes of less diameter than 3 inches cannot be recommended for use in ordinary cases.

331. The subject of house-service pipes demands a

brief notice. The material of which they are made is either wrought iron or lead—the former kind being the stronger and cheaper, but the latter the more durable, because less liable to rapid corrosion. They are generally specified to be of a certain weight per lineal yard ; depending upon their diameter and the pressure to which they will be exposed in use.

The following table indicates the ruling practice.

WEIGHTS OF PIPES FOR HOUSE-SERVICE.

Drawn Wrought-iron Pipes.

Diameter of Bore.	Head.	Weight per lineal yard.
$\frac{3}{4}$ inch	150 feet	$3\frac{1}{2}$ lbs.
1 ,,	,,	$5\frac{1}{2}$,,
$1\frac{1}{4}$,,	,,	7 ,,
$1\frac{1}{2}$,,	,,	9 ,,

Lead Pipes.

Diameter of Bore.	Head.	Weight per lineal yard.	Head.	Weight per lineal yard.
$\frac{3}{8}$ inch	150 feet	$4\frac{1}{2}$ lbs.	300 feet	5 lbs.
$\frac{1}{2}$,,	,,	6 ,,	,,	7 ,,
$\frac{5}{8}$,,	,,	7 ,,	,,	9 ,,
$\frac{3}{4}$,,	,,	8 ,,	,,	11 ,,
1 ,,	,,	12 ,,	,,	15 ,,
$1\frac{1}{4}$,,	,,	14 ,,	,,	20 ,,
$1\frac{1}{2}$,,	,,	18 ,,	,,	26 ,,

332. The joints of iron service-pipes are best formed by short couplings, threaded internally and screwed over the ends of the pipes to be united. The thread on the pipe-ends is cut so as to leave them slightly conical; so that

when they are forced hard into the coupling, with a smearing of red lead on them, a tight joint is ensured.

There are several kinds of joints suitable for lead pipes. The wiped solder joint is perhaps the most reliable. To form this, the end of one of the pipes is widened with a turn-pin, so as to allow the introduction of the filed down end of the other pipe, to which it is to be united. Hot solder is then poured on to the joint, and, whilst in a pasty condition, is wiped around it with a cloth or wisp of tow. This manipulation requires the skill of an experienced plumber.

Fig. 88.   *Fig. 89.*

 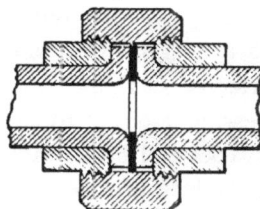

Another plan, which demands less skill, is to widen the ends of the two pipes so as to receive a short brass ferrule. The joint is then enclosed in an iron mould, and hot solder is run in until the space round the ferrule between the ends of the pipes is completely filled with metal fused into the lead of the pipes. The mould is then removed, and a clean, strong, but inflexible joint, *Fig. 88*, results.

The ordinary " union " joint, *Fig. 89*, hardly requires description. The ends of the pipe, slightly up-set, are drawn tightly together with a leather washer between them by screwing up the nut.

333. For small aqueducts, stoneware pipes are often found useful, especially for conduits that follow the contour of the ground with a slight fall ; they can only sustain light

pressures. By employing them, a watercourse may be
formed that is both cheap and free from all risk of pollution, and is superior to open ditches or conduits, inasmuch
as it may be siphoned across small depressions in the
ground—thus frequently effecting considerable saving in
distance and in fall. Such an aqueduct must, however, be
always kept running freely; or it must be provided with
balancing-tanks (§ 357) at such intervals as may be necessary to prevent the occurrence, at any point, of a hydrostatic pressure greater than that due to about 20 feet head
of water.

Stoneware pipes for this work are made from finely-
ground clay, washed, tempered and pugged, and baked to
formed a tough, vitreous, homogeneous mass. They must
be perfectly straight, true in circular section, uniform in
thickness, and ought to exhibit a finely-glazed surface free
from cracks and other flaws.

The joints are generally of the ordinary spigot-and-
socket form, and are made by forcing stiff Portland cement
mortar hard into the annular spaces between the sockets
and corresponding spigots, when the pipes are laid. It is
essential that the cement be used in a stiff pasty condition
and be well rammed into the joints. Soft mortar merely
applied with the trowel does not give a satisfactory result.

During the process of jointing, care must be used to
prevent any cement that may be forced past the back of
the sockets, from being left in the pipes ; where it would
form a serious obstruction to the flow of the water, and
would probably become the nucleus of a complete stoppage
in the future.

It is only by unremitting attention to such details of
execution as this, that aqueducts can be made to fulfil in
practice the conditions contemplated in their design ; and
a systematic consideration of the special operations be-

longing to this class of work, is quite as important as the elaboration of the hydraulic principles involved in it.

334. The construction of those aqueducts that are mere channels lined with clay, concrete, brick or stone, is a matter that requires the exercise of so little special art as to render it superfluous to dwell thereon in a treatise pretending to discuss operations belonging peculiarly to hydraulic engineering. Even timber, and iron or steel conduits, seldom involve in their construction any features distinct from those met with in every-day mechanical work. The building-materials used for this purpose must not be porous, friable, or liable to erosion by the action of the atmosphere or of running water to any considerable extent ; and, in all mason-work, hydraulic lime or cement must be employed.

The formation of aqueducts may involve the erection of such vast structures as that which carries the Marseilles canal across the valley of the river Arc, at Roquefavour ; but it cannot be said that there is anything attaching particularly to waterworks construction in any part of that magnificent pile of masonry, beyond the channel it supports. Again, the piercing of aqueduct-tunnels differs in no respect from tunnelling for any other engineering purpose ; except that the inverts, up to the water-line, must be rendered impervious to moisture, and must be of the proper section for the efficient conveyance of the water through them.

Both tunnels and conduits must of course follow the hydraulic gradient, which, in the case of uniform flow, traverses the surface of the current ; angles and sharp curves in the route should be avoided ; and, in the case of unlined tunnels, a sufficient increment of sectional area should be provided, to counteract the great retardation of flow caused by the ragged surface of the rock ; waste-weirs and sluices are placed at the entrances of tunnels and of long culverts, as well as at the upper ends of siphons and at other convenient places along the line of aqueduct (§ 319). The work involved in all this is generally of a simple character, common to almost every class of engineering building.

335. But when we turn to the wider subject of the conveyance of water under pressure in metal pipes; there is much to be said that is specially applicable to that matter. The assembling, laying and jointing of the pipes and their accessories which are comprised in a large water-main, is a work that requires great labour and considerable manipulative skill. And when these operations are conducted in a rough country, where roads, if they exist, are often unserviceable, the work may well rank in point of difficulty with other classes of constructive engineering which produce a more imposing result.

The weight of a single cast-iron pipe, 40 inches in diameter, ranges between $2\frac{1}{2}$ tons and 5 tons; and these heavy but fragile masses of iron have to be conveyed with the utmost care, often for long distances across hilly country, to their final destination. Where roads are available, they are used as far as possible—the pipes being carried by specially contrived vehicles from the railway-station or the wharf to the nearest point of the pipe-line. If the roads are smooth and level, and the loads are not too heavy, two-wheeled " whims," from the arched axles of which the pipe is suspended in sling-chains, are found useful. A loading crane is thereby dispensed with—the pipes being raised in the slings off the ground by means of capstan-headed screws working in the crown of the arched axles of the whims, and being lowered again by the same means at the place of their deposit.

For heavy pipes, a four-wheeled carriage is the best,

and few contrivances for transporting pipes by road, serve the object better than a short timber-carriage, upon the raised bolsters of which the pipe rests in blocks hollowed to fit the curvature of its barrel.   For crossing soft ground, broad-wheeled vehicles are necessary ; but even with these, the ground alongside of a pipe-trench becomes so cut up that, in wet weather, it would often be impassable for any ordinary wheeled conveyance.   On this account, a tramway is not infrequently laid beside the pipe-track, along which the pipes and other materials are expeditiously transported on trolleys, with a minimum haulage-effort.   Moreover, this tramway may also carry the crane that is employed for laying the pipes in the trench—a method found to answer well in practice.   A 3-foot gauge tramway, of 30-lbs. or 40-lbs. flat-footed rails spiked to wooden sleepers will answer for the heavier classes of pipes ; whilst a light 18-inch tramway, on the Decauville or some similar system, suits smaller sizes of main, less than, say, 2 feet in diameter.

336. In describing briefly the operations of pipe-laying, we allude particularly to mains of large diameter ; but the description, with slight modification, is equally applicable to all classes of pipes.

The pipe-trench is excavated with vertical sides to such a depth as will give a minimum cover, in this country, of 2½ feet of earth over the main, when laid, and of such a width as to allow 6 inches or 9 inches clearance between the pipes and the sides of the trench.

For lead-jointed or flanged pipes, chambers are excavated opposite every joint, in order to facilitate the work of "setting" or screwing up.   These joint-holes are recessed 10 inches to 12 inches into the sides and bottom of the trench.   Care must be taken to set them out from the actual net measurements of the pipes about to be laid ;

otherwise the apparently insignificant differences in the lengths, are found to produce, in the aggregate, a considerable error in work; and the joint-holes, if excavated in wrong positions, have to be either extended or re-cut. This cannot be done without seriously impairing the soundness of the pipe-beds. In wet, or otherwise bad ground, where close timbering is required, it is well to excavate the entire trench of sufficient width to allow the jointers to work in it. Where the excavation is in rock, the trench should be sunk sufficiently low to admit of a 6-inch bed of earth being laid on the bottom for the pipes to rest upon. In soft or boggy ground, the main must be supported upon piles or piers carried down far enough to obtain a solid bearing. The bottom of the trench must be set out and carefully dressed to the proper profile of the pipe-line, so that the pipes may rest upon a fair bed throughout their bearing, without packing. Joint-holes are not required for turned-and-bored pipes—a decided advantage, for they are always troublesome to construct, and they interfere with the bedding of the main.

337. The operation of laying the pipes in the trench is accomplished by either sheer-legs or travelling cranes. When the former apparatus is employed, the pipe to be laid is rolled upon a couple of timber balks laid across the pipe-trench; the shear-legs, one pair of which are united together and carry a double-purchase winch, are straddled across the trench and pipe; a sling-chain is passed round the pipe-barrel and attached to overhead chain-tackle carried by the legs and worked from the winch. The pipe is then raised, its soundness is tested, and it is lowered into position in the trench—the spigot being directed by the pipe-layer into the socket of the last-laid pipe—being first of all, however, used as a "ram" or "mall" to gently force it home into the one that precedes it. In this opera-

tion care must be exercised to avoid splitting sockets or doing other injury ; for which reason a plank is interposed between the swinging pipe and the socket of the one that is to be driven.   The spigot of one pipe should not be forced so far into that preceding it, as to bear hard against the back of its socket.   Copper or lead distance-strips should be inserted in order to prevent such an occurrence, being removed before the joints are made.

When laying a turned-and-bored main, the pipes are generally driven home with a swinging wooden mall.

Where a tramway is laid alongside the pipe-trench, a steam travelling-crane, using one of the tramway rails and a temporary rail set to a wider gauge, may be employed for pipe-laying with advantage.   The labour of rolling the pipes and moving the balks is saved by the jibbing-action of the crane, and the speed of working is greatly accelerated.   Even where a tramway is not laid, a few lengths of metals secured to cross-sleepers may carry a locomotive crane, which, moving its own road forward in sections, will be found to save time, labour and expense.   In hilly and rough country, such a machine, weighing, perhaps, 12 tons or 15 tons, may prove sometimes to be a troublesome piece of apparatus to move about ; and it should therefore be designed with a view to its being readily dismantled and re-erected when occasion requires it.

In laying small pipes, it is usual to joint a number together on the side of the trench, and then, passing slings under them—carefully avoiding any deflection at the joints —to lift and place them in the trench by manual labour. This saves the expense of making many joint-holes, and ensures better work than can be easily obtained in the inconvenient positions required for caulking joints in small trenches.

2 A

As soon as the pipes are laid and adjusted to their positions, they are firmly held by being packed with a little earth on each side of the barrels, and are then ready to be jointed.

338. In making a lead joint, the first operation is to force two or three laps of spun-yarn or soft lead wire to the back of the annular space between the spigot and faucet—a uniform thickness being maintained by steel wedges driven into the joint where necessary and kept there until the resistance of the yarn or lead wire holds the spigot in

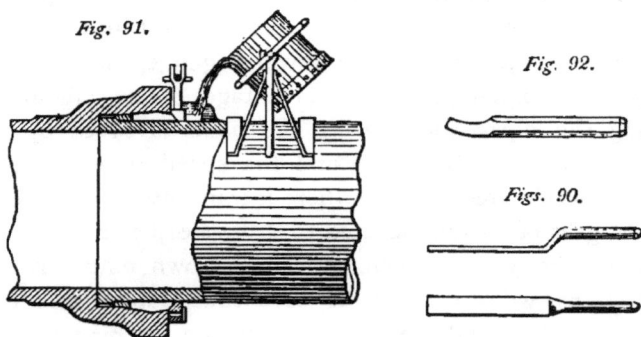

Fig. 91.

Fig. 92.

Figs. 90.

place. If lead wire is used, it permits the adoption of a shallower socket than does yarn—thereby reducing the weight of metal in the pipe.

The yarning-iron, *Figs. 90*, is marked to indicate when the proper depth for the lead joint is obtained. The face of the joint is then closed by an iron or steel ring, *Fig. 91*, hinged at the bottom and drawn together at the top, within 4 inches, by a set-screw, and formed so as to leave a triangular space all round its inner edge. The ring is greased and made tight against the pipe by clay luting round its edges—the open space at the top being formed into a small cup by a piece of clay. The lead-pot, containing a sufficient quantity of molten lead, is then set on a saddle placed across the pipe, and is tilted by the handles on the ends of its trunnions so as to pour the molten metal into the cup, whence it flows into and around the joint, completely filling it. The ring may be then removed and the joint is ready to be " caulked " or " set up." In order to avoid the occurrence of air-holes and other flaws in the joint, both the spigot and the socket must be perfectly dry and clean, and the "running" or pouring of the lead must be rapid and continuous. With a view to facilitate the complete escape of air during this process, a small hole has sometimes been drilled through the back of every socket—a precaution involving some practical difficulties in laying the pipes, and, according to our experience, unnecessary. It is, however, desirable that the pipes should not be laid with the sockets facing down-hill on gradients steeper than 1 in 24 ; and the best practice is, wherever possible, to lay up-hill from the scour-valves (§ 351).

The lead used must be pure, but not too soft. Highly de-silverised lead may often be improved for use in joints by the addition of a little tin to harden it. The temperature of the molten metal requires attention ; if too cold, it may fail to run round the joint, and if too hot, it is apt to set hard and become liable to fracture under the application of the caulking-set.

After the jointing-ring has been removed, the "git," or solid contents of the cup into which the lead is poured, is cut off with a sharp chisel ; with which tool also, the edge of the lead fillet left by the jointing-ring is gently loosened from the pipe-barrel. The jointer then proceeds to caulk the lead into the joint, in three successive operations, with an equal number of properly-proportioned " sets," *Fig. 92.*

339. The caulking-sets should be $\frac{1}{8}$ inch, $\frac{3}{16}$ inch, and $\frac{1}{4}$ inch thick at the face, respectively, and $\frac{3}{4}$ inch or 1 inch

broad, according to the size of the pipes they are used upon. They are directed by the jointer, and are struck with a double-handed hammer of 5 lbs. weight by his mate. Each stage of the caulking-operation should be proceeded with continuously round the joint ; the surplus of the fillet of lead which is not forced into it, being gradually squeezed up in a complete ring, until it is sheared away from it by the sharp edge of the inner periphery of the socket. With small pipes, a clay rope may be substituted for the jointing-ring, and the caulking is usually performed entirely by the jointer with a single-handed hammer. During the process of jointing, and subsequently until the pipes are covered with earth, they should be screened from powerful solar rays, which are apt to produce distortion and drawing of the joints, especially in the case of mains of large diameter.

340. The quantity of lead used in forming joints is very variable, with a tendency to excess unless strict supervision is exercised over the work. As it forms an important item in the cost of mains we give the undermentioned weights of lead joints, $2\frac{1}{2}$ inches deep, ascertained by weighing large numbers of them of ordinary thickness :

Diameter of pipe	..	36 inches.	18 inches.	9 inches.	4 inches.
Lead in the joint	..	50 lbs.	21 lbs.	10 lbs.	5 lbs.

Much, of course, depends upon the thickness assigned to the joints and the accuracy observed in casting the pipes ; but these weights may serve to indicate what is found in ordinary practice.

341. Two methods of uniting adjacent lengths of main may be noticed : (1) by the insertion of one or two double-spigot pipes, *Figs. 93*, cut so that the spigots may be about 6 inches apart, with a collar-joint. The collar, or "thimble" as it is sometimes termed, is passed on to the first-laid

spigot, and is subsequently slid over the opening, when a lead joint is made at each end of it. (2) A method that is only applicable to small pipes, is to cut the spigot end of the last pipe to be laid to such a length that it will just fit in its place ; then to raise several pipes on each side of the junction into an arch-form, so as to allow of the introduction of the middle spigot into its corresponding socket ; after which the pipes are again lowered on to their beds and the union is completed by the jointers.

Figs. 93.

SOCKETS FOLLOWING

SOCKETS MEETING

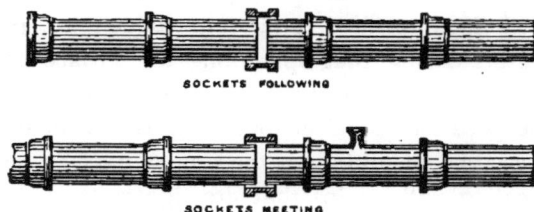

342. The joints of turned-and-bored pipes are very simply made. The socket of the last-laid pipe is heated, generally by a fire-basket suspended within it, to a temperature of about 250° F. ; the turned portion of the spigot of the pipe to be next laid, is brushed over with hot asphalt or with cement, a few laps of twine being sometimes added, and is thrust into the hot socket and driven firmly home by a few blows from the swinging mall.

343. Flange-joints, if accurately machine-faced, are made with red-lead putty, a fine strand of hemp laid once or twice round the joint being sometimes a serviceable addition to it, and are drawn firmly and evenly together by the bolts and nuts. An excellent joint, requiring less accuracy of workmanship than the former, may, if the flanges are planed right across, be made by inserting a canvas disk, properly formed for the bolts to pass through

it and soaked in a thick mixture of red lead and boiled oil, between the meeting faces, which are then duly drawn up. Unplaned flanged joints may be made with a complete ring of $\frac{3}{8}$-inch lead-wire inserted within the line of the bolt-holes and squeezed flat between the flanges by the pressure of the bolts.

344. The final operation of filling-in the pipe-trench is more important than might appear at first sight. The excavated earth is restored and carefully packed in thin layers under the pipe, in such a manner as to afford it a firm bearing throughout its length. The upper part of the trench is then filled, and well trodden or "punned" with beaters during that operation—the surface being restored as nearly as may be to its original condition ; for which purpose the mould or road-material, as the case may be, is preserved separately at the time of excavation.

In dealing with mains that are lighter than the water they displace, care must be used to ensure a dry trench whilst the pipes are being laid, and the filling-in must proceed down-hill, so as not to impound water in any part of the pipe-trench. A pipe, 40 inches in diameter and 1 inch thick weighs 46 cwt., and displaces 66 cwt. of water. If closed at the ends and immersed in water, or even in semi-fluid earth (slurry), every 12-feet length would need to be loaded with at least 1 ton to prevent it from floating. This is precisely the condition of a portion of an empty water-main, when the pipe-trench has been filled-in at two points and water is impounded between them.

345. Turned-and-bored joints produce an inflexible main. Where lead-jointed pipe-lines follow a slight curve, ordinary straight pipes may be used ; but the permissible curvature is limited by the depth of their sockets and the thickness of the joints. Lead joints properly coned or swelled in section, *Figs. 78, 79,* will permit without failure,

a slight curvature of the main to take place after it has been completed, such as that which sometimes arises from unequal settlement of the pipe-beds. This does not, however, apply when the curvature is considerable. In situations where the curvature of a main is considerable, irregular, and liable to alteration, special flexible-jointed pipes are introduced. *Figs. 94* and *95* illustrate two kinds

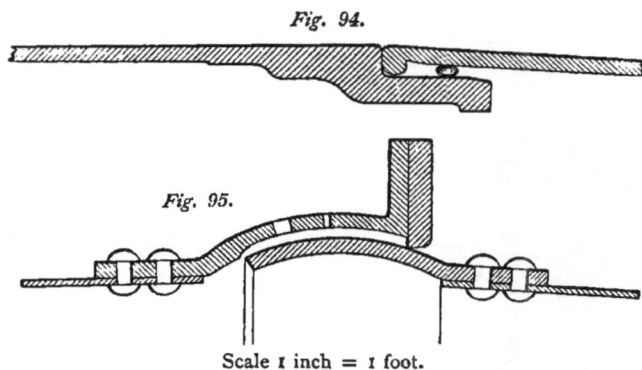

Fig. 94.

Fig. 95.

Scale 1 inch = 1 foot.

of these joints ; the former has been used on several works in passing mains through soft ground, and the latter was recently employed successfully in crossing the bed of the river Mersey.

In large mains, the normal thickness of joint may be taken at $\frac{5}{16}$ inch, and the minimum at $\frac{3}{16}$ inch—the depth of the socket being 5 inches. The maximum deviation of one pipe from the axial line of its fellow is, therefore, $\frac{A B}{O A} = \frac{1}{8}$ inch in 5 inches, *Fig. 96*, i. e. 0·3 foot in a 12-feet pipe. The chords of the curve round which the pipes are laid with this deviation, are 12 feet long ; and the curve is therefore circular, and its diameter B C, *Fig. 97*, is

$$\frac{A O . A A_1}{A B} = \frac{12 \times 24}{0 \cdot 3} = 960 \text{ feet.}$$

The minimum radius at which ordinary straight pipes may, under the conditions above stated, be laid, is then 480 feet. If a curve of less radius than this must be followed, shorter lengths, or specially-curved pipes are used in good work ; although it is notorious that straight pipes are often forced round curves to which they are quite inapplicable, with the inevitable result of bad joints.

Fig. 96.   *Fig. 97.*

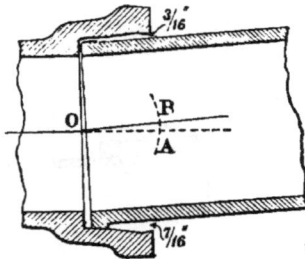
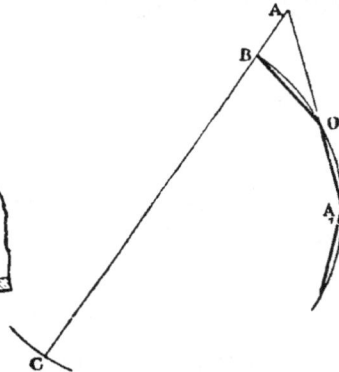

346. Bend-pipes are specified according to the situation they are designed to occupy, as of certain radius and length, or of certain angular deviation—the latter description being generally applied to those bends of small radius which are often required in pipe-lines that follow highways and other narrow routes. When a main is charged with water there must always be a certain hydrostatic pressure tending to force the joints of a bend. This pressure is the resultant of two equal forces acting normally to the planes of the joints under consideration, and of amount equal to the product of the sectional area of the pipe into the hydrostatic pressure.

For example, in a 15-inch main, the static thrust on a

60°-bend, acting in the direction of its middle radius, is, if the pressure be, say, 200 lbs. per square inch,

$$2 \times 176 \times \cos 60° \times 200$$
$$= 35,200 \; lbs.$$

It is evident that this pressure, exceeding 15 tons, must be opposed by some artificial resistance superior to that afforded by the rigidity of the joints and the soft earth-filling of the pipe-trench. The bend is consequently backed by a mass of concrete, bearing solidly against the side of the trench, which effectually prevents the joints from being drawn by a forward movement of the pipe.

In addition to hydrostatic pressure, bends are subject to a radial thrust, due to the motion of water through them. Its amount is in general small, as the following investigation indicates :

Consider the case of steady laminar flow through a circular bend of radius r and sectional area A, the angle at the centre of which is θ, *Fig. 98.*

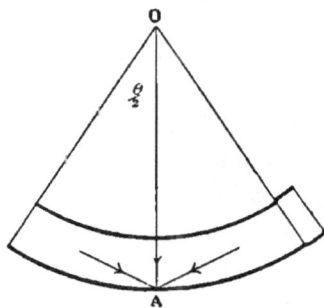

Fig. 98.

Let v, as usual, denote the velocity of the flow of any stream-line, and w the weight of unit-volume of water. The resistance to the impact of the water upon an elementary ring of the pipe situated at a distance $r\phi$ from A along the bend (ϕ being a variable angle), in a direction parallel to the line O A, which bisects the angle θ, measured by its centrifugal force, is

$$\frac{A \, w \, v^2 \, r \, d\phi}{g \, r} \cos \phi \, .$$

The total resultant thrust exerted upon the bend in the direction O A, is

$$\frac{A\,w\,v^2}{g} \int_{-\frac{\theta}{2}}^{+\frac{\theta}{2}} \cos \phi \, d\phi ,$$

$$= \frac{2\,A\,w\,v^2}{g}\, \sin \frac{\theta}{2} .$$

This may be of importance in a short pumping-main where it forms a sensible fraction of the entire pressure dealt with ; but is, in mains of considerable length, so much less than the head lost by friction when the water is in motion that it may be neglected as a practical consideration.

It may be pointed out that the foregoing considerations, whilst showing generally the state of affairs under the conditions assumed, are not entirely satisfactory. Further, no rational data exist by which the head lost at bends may be computed precisely. There is no doubt that the eddies caused by the radial pressures developed in the flow round a pipe-bend, or a bend in any kind of aqueduct, do absorb hydraulic energy to a certain extent ; but from observation and experience we do not consider that the head so lost at a single bend of even small radius is in general a sensible quantity.

347. The effects of impact arising from the sudden arrest of the motion of the column of water conveyed in a long main, are well worthy of consideration, and might be expected to influence, in an important degree, the design of pipes. The momentum thus lost by the water is expended partly in distending the main and partly in effecting compression of its own bulk.* Here again, we must confess

* The difficulty of separating the compression of water from that of the vessel containing it, forms an obstacle to its precise determination. For

our inability to state in statical measure the amount of the forces thus developed, from a knowledge of the facts as they are ordinarily ascertained. Expressions, indeed, have been put forward, derived by certain mathematical processes, pretending to give the force of impact, or "water-ram" as it is sometimes called, resulting from the rapid closing of a valve on a long water-main. But, as a matter of fact, such forces can only be expressed in statical measure if the time-rate of the process of arrest is accurately known; and this infinitesimal element being, from the circumstances of the case, impossible to observe— the problem is not amenable to mathematical treatment. A better notion of what may be expected to occur can be derived from considering the results of some actual experience.

In 1884, Mr. E. B. Weston described a series of elaborate experiments made by him at Providence, R.I., to determine the increase of pressure in a water-pipe consequent upon the rapid checking of the flow through it.* The trials were conducted upon a length of 30 yards of 6-inch pipe, branching off a 24-inch main, followed by 22 yards of 4 inch, 1 yard of $2\frac{1}{2}$-inch, 2 yards of $1\frac{1}{2}$-inch and 2 yards of 1-inch pipe. The water was caused to flow through this system of piping, its mean velocity in each section being computed from the measured delivery; and the flow was then checked by closing an outlet-tap provided for the purpose—an operation that occupied 0·16 second. The water-pressures developed by thus abruptly

---

practical purposes, the compressibility of water, as determined by experiments in which the envelope was of copper, may be stated. It is 0·000047709 (approximately $\frac{1}{21000}$) per standard atmosphere (760 mm.).

Regnault, 'Relation des expériences entreprises pour déterminer les principales lois et les données numériques qui entrent dans le calcul des machines à vapeur,' tome i. p. 455—Paris, 1847.

* Trans. American Soc. C.E., vol. xiv p. 238.

stopping the current, were measured by a Richards steam-
indicator, which was tested both before and after use. The
following are sufficient to exemplify the results obtained:

Diameter of pipe.	Velocity, feet per sec.	Ram, lbs. per sq. inch.	Velocity, feet per sec.	Ram, lbs. per sq. inch.
1 inch	10·02	106·3	20·94	177·5
1½ ,,	4·45	81·8	9·31	121·5
2½ ,,	1·60	52·0	3·35	99·0
6 ,,	0·28	15·8	0·58	36·8

In an experiment published by Mr. A. R. Binnie, in
1887,[*] a ¾-inch pipe 114 feet long branching off a supply-
main was furnished at the end with a plug-cock, the effective
waterway of which was 0·152 square inch.

The following pressures were indicated by gauges;

	At the branch. Lbs. per sq. inch.	At the cock. Lbs. per sq. inch.
Cock not open	125	125
,, open	120	20
,, shut quickly	220	550

These pressures are astonishing, but although highly
suggestive, we cannot, in the absence of a statement of the
velocities, deduce exact rules from them.

Mr. Weston's experiments point to two definite conclusions: (1) that the pressures due to hydraulic impact
vary as a lower power than the square of the velocity;
(2) that, with the same velocity the intensity of pressure
developed is higher in large than in small pipes. Although

* 'Lectures on Water-Supply'—Chatham.

the data here presented are insufficient to lead to the construction of a general law applicable to the subject, it is evident by analogy from all these observations, that, under similar circumstances, exceedingly heavy blows might be expected in large mains ; and that the ample margin of strength adopted for water-pipes is entirely justified by the extraordinary pressures that may be anticipated to occur sometimes in practice.

In order to avoid water-ram in large pipes, when it is necessary to stop the flow, a valve is shut as near to the upper end of the works as may be, consistently with its situation being such that no portion of the main on its lower side is above the level of the reservoir next below it. Moreover, all valves are constructed so that rapid closing is impossible ; and, as a further safe-guard, relief-valves (§ 354) are placed in proper positions upon the main.

348. In constructing a pipe-siphon across a steep-sided valley, the inclination of the main may be so high, that its natural adhesion to the ground becomes insufficient to prevent a considerable longitudinal stress from being transmitted from pipe to pipe, to the risk of fracturing the sockets. This stress may be so great as to make it advisable to support some of the pipes in such a manner as to transmit the thrust they receive from those above them, directly to the ground. Where the work is in rock, this may be effected by building the sockets of such pipes into a solid mass of concrete, keyed into the walls and floor of the pipe-trench. But if the sides of the valley do not lend themselves to this method of tying down the pipes, masonry or concrete anchors are built at some depth below the surface of the ground ; and through them are passed steel rods, attached at one end to trunnions cast on the pipes to be supported, and at the other end made fast by cotters

to iron plates bearing against the anchors. In any such case, brackets inside the sockets are necessary, and must be dressed to provide an even bearing for the spigot that rests upon them. A bearing-ring of hard lead is sometimes inserted between the brackets and the spigot.

349. When pipes are laid through deep trenches, the great vertical load imposed by the settlement of the earth, and the slight lateral support to the barrel that can be afforded by the filled-in earth, must be noticed. This does not apply to cases of tubes carried through semi-fluid earth which transmits a considerable lateral pressure, and does not produce the distorting effects that are caused by a load of dry earth. A cast-iron pipe, 40 inches in diameter and 1 inch thick, laid in a hard-bottomed trench filled-in with dry material, will not carry more than 12 feet depth of earth without distorting, no matter how well the barrel may be supported by ordinary earth-packing at the sides. With twice that depth of superincumbent soil, the distortion of such pipes has, within our experience, amounted to as much as $\frac{5}{8}$ inch of vertical diminution and horizontal increase of diameter—an amount that approaches the maximum distortion the pipes can suffer without fracture.

No reliable rule can be given for designing the dimensions of pipes under conditions like these. A judgment based upon the observed effects of similar phenomena is the most trustworthy guide.

From the foregoing allusions to one or two of the practical difficulties that beset the construction of a water-main, we pass to a brief review of the auxiliary apparatus employed therein.

350. A water-pipe, following the natural undulation of the country it traverses, must be provided at every summit with means for the escape of air that is enclosed in it

previous to its being charged, and is continually disengaged from the water.

If this provision were omitted, charging fully would be almost impossible; and, even if accomplished, the summit-portion of the main, *a b, Fig. 99,* would gradually become filled with air—the point *b* being situated at such a level with respect to the reservoir B, that the pneumatic pressure there would be sufficient to cause a certain definite flow of water through the lower part of the pipe into the reservoir B—that water being supplied by overflow at the summit *a.*

Fig. 99.

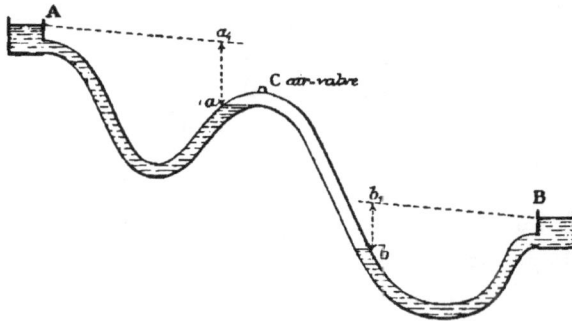

The precise amount of the flow is determined by an imaginary hydraulic gradient from A to the point *a,* where $a\,a_1 = b\,b_1$ is the head of water equivalent to the pneumatic pressure obtaining at the summit of the main.

Serious diminution of the flow would take place long before the extreme condition illustrated by the diagram *Fig. 99,* was reached; but the action may be entirely prevented by inserting an air-valve at C, which will permit the escape of air, though not that of water, out of the main. *Fig. 100* shows a type of automatic air-valve in common use. An indiarubber ball rests in the position indicated, if there is air in the pipe, and remains there until all of it has

escaped through the small orifice above it.   As soon as the
water which has expelled the air, completely occupies its
place, the ball floats and is pressed against a dished seat,
thus closing the orifice.   There are other valves in use for
this purpose, though none, perhaps,
simpler and more effective.   The indiarubber balls require to be renewed from
time to time.

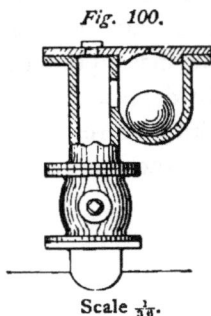

Fig. 100.

Scale $\frac{1}{30}$.

The size of the orifices in air-valves
is an important feature.   If they are
too large, the escape of air during the
process of charging the main, may be
so rapid as to give rise to dangerous
concussion of the water in the pipes ;
whereas, if they are very small, the
charging, after the main has been emptied at any time for
repair, may be unduly prolonged.   Their diameters generally range between $\frac{1}{8}$ inch and $\frac{3}{8}$ inch.

It is useful to have a gun-metal screw-plug of comparatively large diameter in the flanged cover of the valve ; this
may be taken out whilst the main is being charged and
is replaced as soon as water begins to issue through the
hole.   The automatic valve may, in that case, conveniently
have a fine orifice, for regulating the subsequent discharge
of air disengaged from the water during ordinary working.
The air-valve casting is attached to the flange of the
special branch-pipe ; a plug-cock being inserted between
the two, in order that the water may be shut off the air-
valve when it is necessary to open it for inspection or
repair.   As mains are in general laid up-hill, two sections
ordinarily meet at a summit, where an air-valve special is
introduced and union of the two is effected by means of a
double-spigot pipe and collar, as illustrated in the lower of
Figs. 93.

351. At the bottom of every hollow in the water-main, a scour-valve should be provided, both for scouring out, when necessary, the sediment that is invariably deposited at those points, and for emptying the main. As a general rule, the scour-valve may be made with a diameter one-fourth part of that of the main it serves. In construction it is similar to the stop-valves described hereafter. It is generally necessary to lay a few lengths of pipe from the scour-valves to convey the flush of water into the nearest available water-course ; and, sometimes, these waste-pipes are of considerable length. The flanged branch to which the scour-valve is bolted forms, like the air-valve branch, part of a special pipe. In this case, how-ever, the special is a double-socket pipe ; from which as a starting-point, the main is laid in opposite directions. *Fig. 101* illustrates diagrammatically a scour-valve attached to a main by a flanged bend.

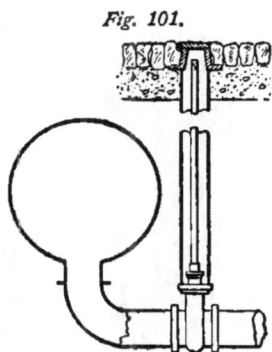

Fig. 101.

The bolts connecting such bends and the scour-valve to the main pipe, must be designed to sustain tension equal to the pressure in the main upon the area of the scour-branch.

Sometimes, and very conveniently in the case of small pipes, the branch is brought tangentially from the bottom of the main ; so as to empty it completely, and, at the same time, to keep the level of the waste-pipe as high as possible.

Another plan by which depth may be saved, is to incline the branch at about 30° with the vertical, adding a 60°-bend to bring the scour-valve and waste-pipe horizontal.

2 B

This is often of importance in flat country, where the absence of a water-course at a convenient level, though it may not prevent the scouring of the main under pressure, makes it sometimes necessary to empty the lower portion of it, when required, by pumping.

352. At certain points along the pipe-line, stop-valves are placed to regulate the rate of supply, and so that the sections of main between them may be isolated for repair.

Fig. 102.

The former object is best effected as near as possible to the outlet of any portion of the main, i. e. where it delivers into a reservoir; but for the latter purpose, such stop-valves are conveniently situated at depressions in the pipe-line, and are provided in every case with a scour-valve on each side of them. The mechanical details of these valves are not within the scope of this treatise; but a general idea of the apparatus is presented by *Fig. 102*, which illustrates an ordinary type of double-faced valve. The shutter is actuated by a powerful screw of fine pitch. All the working parts, faces, glands and bearings are of gun-metal. It is customary to reduce the area of the water-way of stop-valves upon large mains considerably below that of the pipes commanded by them—the diminished thrust against the valve-shutter when closed, and the consequent ease of opening it, being gained without sensible loss of energy in the main when the valve is fully open, such as would result from an enlargement of the area of the pipe. A small rider-valve is also provided, which being opened

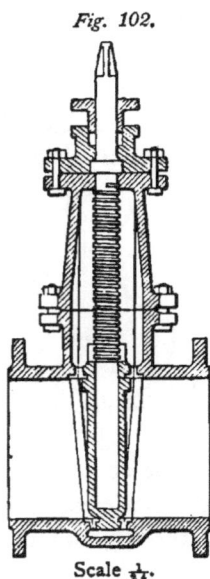

Scale $\frac{1}{11}$.

before, and closed after, the main valve, renders the manipulation of the latter comparatively easy ; its effect being to
prevent excessive difference of pressure upon the two faces
of the main shutter when the flow is checked, and to
charge and equalise the pressure in a section that has been
emptied, when it is required to put the pipe in train with
that on the other side of the stop-valve.

Fig. 103 shows a view of a 44-inch stop-valve with three
shutters, made for the Liverpool waterworks.   In this

Fig. 103.

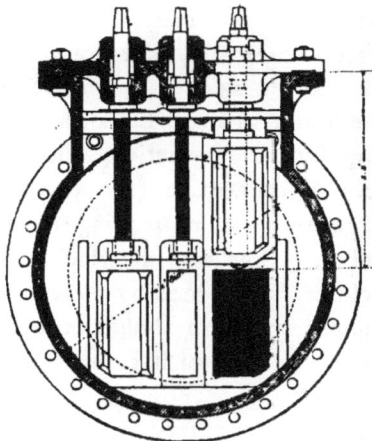

Scale $\frac{1}{18}$.

valve, the middle shutter closes the rider-valve and is much
smaller than those on either side of it.

The fracture of mains that have been working for some
time, is almost invariably due to hasty closing of stop- or
sluice-valves.   And the necessity for operating them very
gradually, particularly towards the end of their stroke,
cannot be too forcibly impressed upon those whose duty it
is to manipulate them.

2 B 2

353. Automatic stop-valves are provided where serious damage or inconvenience would attend the discharge of water due to accidental fracture of the main.

They are placed, if it is feasible, under comparatively light pressure—being often situated a little below the hydraulic gradient, on the verge of deep depressions in the pipe-line, where serious fractures are chiefly to be apprehended. *Fig. 104* shows in outline the ordinary form of these valves. A pendulum, dotted in the *Fig.*, carries a small disk within the pipe, normal to its axis, and is balanced by a weight and adjusted so as to hang vertically when the velocity in the main is normal. Should it, however, increase abnormally, as it would in the event of a fracture on the lower side of the valve, the additional pressure of the water against the disk deflects the pendulum, releasing, by a trigger-action, a heavy weight which closes the main throttle-valve (the outline of which is dotted in the figure) by a chain passing round the periphery of the valve-wheel. The rate at which the blade of the throttle-valve closes, is governed by a cataract-action ; and, in a large main, the operation should occupy not less than 15 minutes—the latter part of it being effected at a gradually diminishing rate.

This is ensured by automatic throttling of the cataract supply-pipe, which takes place through the medium of a cam and lever as the main valve closes. A force-pump is provided for the purpose of setting the valve in position, by means of which also it may be worked through a small range occasionally to prevent it from sticking. All the working parts and bearings of the apparatus are gun-metal.

354. On the upper side of the main, in front of the self-closing valve, an equilibrium relief-valve, *Fig. 104,* is placed.

This valve is kept closed against the normal pressure
of the main by a powerful helical spring loaded with a

Fig. 104.

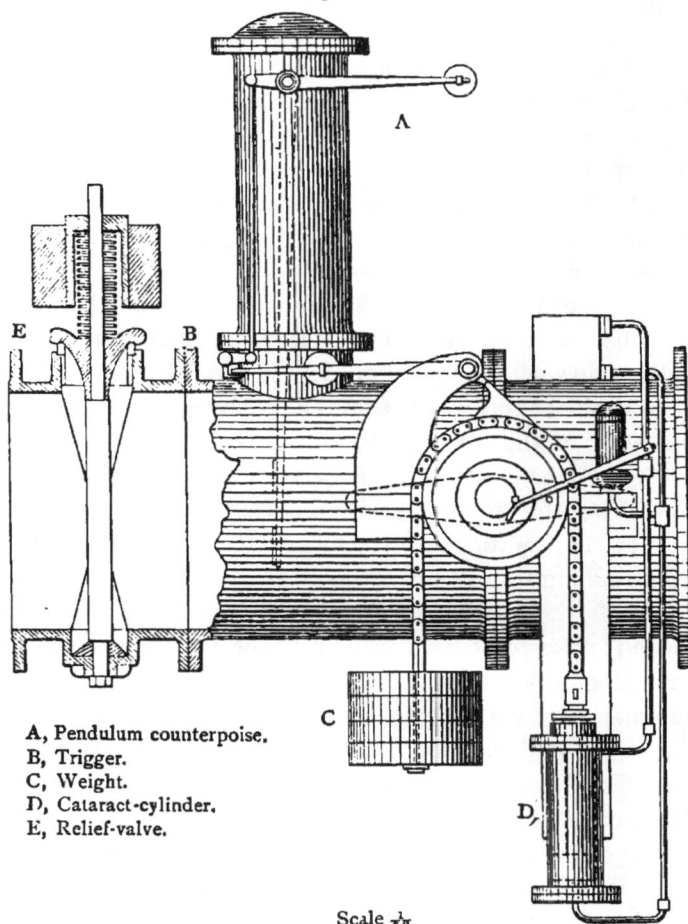

A, Pendulum counterpoise.
B, Trigger.
C, Weight.
D, Cataract-cylinder.
E, Relief-valve.

Scale $\frac{1}{18}$

weight. The inertia of the weight and the stiffness of
the spring prevent the valve from lifting under ordinary
conditions; but any sudden increase of pressure in the

main lifts the valve, compressing the spring, the inertia of which is slight, and permits the escape of a little water, thus relieving the pipes from concussion. By the so-called " equilibrium " method of suspension, a considerable area of discharge around the large valve-disk is afforded by a slight lift of the latter; and the action of the valve, being dependent upon the difference of the areas of the upper and lower disks, may be made very sensitive under the highest pressures.

This form of safety-valve forms an important adjunct to any apparatus the action of which may give rise to rapid arrest of the motion of the water in a main.

The need for automatic throttle-valves of the kind described is sometimes avoided, by arranging that one walks-man shall be always on duty in the neighbourhood of a stop-valve, upon which a pressure-recorder is fixed. In the event of the main bursting, the pressure-recorder falls and rings a bell, attracting the attention of the walks-man, who closes the stop-valve with due caution.

355. In addition to machinery for preventing water from running to waste in the normal direction of flow—the effect of which would ordinarily be to merely cause the waste to take place by over-flow at the service-reservoir or balancing-tank next above it—apparatus is provided to automatically cut off the supply to such reservoirs when the over-flow level is attained by the water in them. A simple self-closing valve in common use for this purpose is illustrated in *Fig. 105*. On the up-turned inlet-pipe is bolted an iron cylinder having a number of vertical equidistant slots in it. Inside the cylinder slides a hollow trunk or thimble, in which are holes corresponding in position with the vertical slots, the lower part being plain so as to cover the slots when it is raised sufficiently.

A vertical spindle attaches the thimble to a float.

When the reservoir is empty, and until the water rises up
to the stop that supports the float under those conditions,
the thimble is in its lowest position and the valve is full
open ; but as the water-level rises higher, the float is lifted,
and with it the thimble, until, at the top-water or over-flow
level, the valve is completely shut.

Further, to provide against the discharge of water
backwards through the main, in the event of fracture

Fig. 105.   *Fig. 106.*

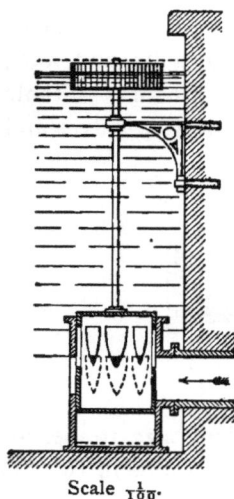
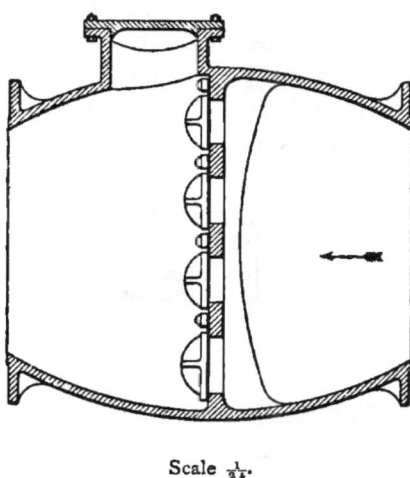

Scale $\frac{1}{100}$.   Scale $\frac{1}{24}$.

occurring in a siphon below the level of the reservoir into
which it delivers, or at some other place where a leak
might establish a reverse flow in the main and lead to the
loss of a large quantity of water, "re-flux" valves are
introduced at such points as the inlets to reservoirs or
towards the "delivery" sides of deep depressions in the
pipe-line, which may be regarded as offering a certain
menace to the works in the respect mentioned. The reflux-
valve, *Fig. 106*, consists simply of a bulb-shaped pipe,

across which there is a strong diaphragm perforated by valve-openings, the collective area of which equals that of the ordinary pipe. These openings are furnished with flap-valves lifting in the direction of normal flow, which automatically close in the event of reversal of the current in the pipe.

Absolute water-tightness is not essential in these valves, but strength and durability are of the highest importance ; it is therefore desirable to make the flaps and the valve-seats entirely of gun-metal.

356. Man-hole pipes, or hatch-boxes, the access-holes being furnished with covers secured by bolts and nuts, *Fig. 107*, are provided at intervals on all mains ; suitable situations for them being on each side of stop-valves and at sharp bends.

Fig. 107.

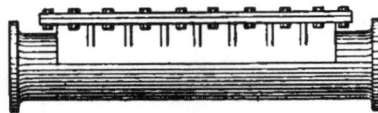

Scale $\frac{1}{24}$.

The hatch-ways must be of sufficient size to permit the introduction and withdrawal of pipe-scrapers (§ 381) through them. In large mains, the hatch-box illustrated in the figure is replaced by a short cylinder, connected at each end to the main pipe by means of collars that are bolted together and can be readily removed when requisite. Main valves and manhole-pipes should be set in brick or masonry chambers, covered either with a flag or with a locked iron surface-box.

357. The last accessory to a gravitation-main which we propose to describe here, is the system of " balancing tanks " and " relieving-reservoirs."

The operation of the former, analogous to that of stand-pipes on pumping-mains (§ 150), is to permit the maintenance of a definite maximum head of water on the section of the main upon its lower, or outlet, side—whilst preventing any increase of that head, by discharging surplus water received into it, through a waste-pipe or over a weir, as the case may be. Suppose, for example, that a supply of water is derived from a source situated at an elevation of 500 feet above the place of distribution, and a head of 250 feet is required for the purposes of that distribution ; whilst a 250-foot head is absorbed by friction in the supply-main under normal conditions of flow. In order to avoid the expense of unnecessarily heavy pipes, and the contingent disadvantages of the inordinately high pressure to which the distribution-system of pipes would be subject when there was little or no draught of water from them ; a "balancing-tank" provided with an overflow would be placed at an elevation of 250 feet. This tank would both act as a service-reservoir if one were needed there, and would, by discharging surplus supply from the source until such time as all the valves were properly regulated, prevent any increase of head beyond the assigned limit of 250 feet on the lower portion of the main and the distribution-pipes.

The "relieving-reservoir," the effect of which is analogous to that of a reducing-valve (§ 362) in a distribution-system, is applied where the available head on a main is so greatly in excess of that required for the conveyance and distribution of the supply, that it is desirable to cause a rupture of the flow, either by a cataract or by throttling the main at its inlet into the reservoir. The relieving-reservoir is in the latter case, in effect, a balancing-tank with the inlet-valve partly closed. The question may naturally arise, why not design the main above the

reservoir of such dimensions as to absorb the surplus head by friction in pipes of smaller diameter, and therefore of less cost ?

One reason for this is that the diameter of a main is limited by the (generally small) permissible velocity of flow through it; and, further, circumstances may arise in which the hydraulic gradient must, from the configuration of the ground between the points of supply and delivery of a main, be considerably less than would be ordinarily inferred from the difference in level of those points—and the lower part of the main would not run full, unless the inlet to the reservoir were throttled.

The dimensions and style of construction of such tanks and reservoirs vary with the magnitude of the supply and local conditions. They may be small brick or concrete wells, iron or steel tanks, or walled or embanked reservoirs. The general principles of their design and construction are identical with those applicable to the several classes of reservoir treated under "the storage of water," Chapter IV.

358. In concluding this chapter, we will allude to one or two of the many special works incidental to the construction of aqueducts of considerable magnitude.

Railways are crossed overhead or underneath according to their situation in deep cutting or otherwise. Where a main is carried overhead, and the span does not exceed 30 or 40 feet, cast-iron flanged pipes may be used, forming a girder and an aqueduct combined ; but if the span is larger, up to, say, 100 feet for a 42-inch main, a riveted steel or iron tube is to be preferred. In either case, the pipe springs off saddles resting upon brick or masonry piers. A slight upward camber is frequently given to such self-supporting built tubes ; but experience shows that it is not necessary to introduce the arch-action into them.

An expansion-joint is provided at one end if the pipe exceeds 50 feet in length; and, unless of very large diameter, it is lapped with felt or other non-conducting material surrounded by wood-lagging, as a protection against frost.

Mains laid under a railway are generally carried through a brick or masonry culvert, of sufficient size to admit of access for inspection and repair; so that fracture of the pipes may not endanger the stability of the line. When the aqueduct consists of a culvert at such a place, it may, nevertheless, be carried under the railway by a pipe within a culvert; by which means certain risks attending settlement or other distortion of brick or masonry watercourses may be avoided, and inspection of the works is facilitated.

A stop-valve should be inserted at each side of the crossing, with a scour-valve between them so that the pipe may be isolated and emptied rapidly if necessary. Crossings of canals and rivulets are similarly dealt with; except that, when these are of insignificant size, the main is frequently laid in a trench excavated in their beds, and is protected by an apron of concrete over them extending a few yards on each side of the pipe-line.

359. An idea of the magnitude and difficulty of the works sometimes necessary where large rivers are crossed, may be conveyed by an account of the recently-completed aqueduct-crossing of the river Mersey, in connection with the new Vyrnwy water-supply to Liverpool. This aqueduct-tunnel is built of cast iron, as also are the shafts. Its length from centre to centre of the shafts is 800 feet, and the depths of the shafts to the invert-level on the Cheshire and Lancashire shores of the river are 46 feet and 52 feet respectively. The external diameter of the tunnel is 10 feet and that of the shafts is 10 feet 9 inches, widened at the bottom to 15 feet. The fall in the tunnel

from the Cheshire to the Lancashire shaft is 6 feet, and a sump is provided at its lower end.

The minimum height of the river-bed above the tunnel is 20 feet.

The shafts are built of flanged cast-iron rings, 4 feet deep, varying in thickness between 1 inch and $1\frac{1}{16}$ inch down to the widened-out part, which is built of smaller segments.

The tunnel is built of rings in segments 18 inches long. In each ring are 10 segments, 3 feet by 1 foot 6 inches and $1\frac{1}{16}$ inches thick, weighing 4 cwt. each, and a key-segment 1 foot by 1 foot 6 inches, weighing 1 cwt. The shafts were sunk by weighting the cylinders and excavating with a grab in the ordinary way. The strata through which the tunnel was driven consisted of layers of clay, sand and gravel of ever-varying thickness. The driving of the tunnel was effected, under air-pressure of about 20 lbs. per square inch, by means of a shield which surrounded the segmental rings, leaving a space of $\frac{3}{4}$ inch clear all round them. This space was filled with clay as the shield advanced, in order to prevent escape of the compressed air. The shield was provided with a steel cutting-edge armed with teeth. At a distance of 2 feet 6 inches from the cutting-edge, a diaphragm was inserted across the shield, the lower middle portion of which was removed. Behind this lower portion of the diaphragm and attached to it round the edges, was fixed a pocket, open at the top, in which the workmen stood to excavate the material that was traversed by the shield.

The lip of the pocket was 6 inches above the centre of the shield. It thus formed a water-trap, so that, in case the water should come into it, the pressure of the air in the tunnel would prevent it from rising higher than the level of the centre of shield.

Round the diaphragm and close to the skin of the shield, were fixed 10 hydraulic rams 7 inches in diameter, working up to 2500 lbs. per square inch ; which, abutting on the ring of the tunnel last fixed, were employed to force the shield forward a sufficient distance, 18 inches, for the next ring to be built in. The shield was long enough when so pushed forward to lap over the last segment built.

As a rule, the ground under the influence of the air-pressure was dry, and the men excavated it standing either in the pocket or, when it was very firm, in front of the diaphragm. In the former case, the pressure was kept on the rams continuously, and, as each spadeful of material was removed, the shield gradually crept forward ; in the latter case, the pressure was not put on the rams until 18 inches of material had been excavated in front of the diaphragm ; after which the shield was driven forward into the space so prepared. The next ring of segments was then fixed ready for the excavation to be resumed as before. Under favourable circumstances, 5 rings of segments were fixed in a day of 20 hours. The spaces between the flanges of the segments were caulked with Portland cement. Through the tunnel, for the first instalment of the supply, a steel pipe 32 inches in diameter was laid, carried on green-heart beams in the tunnel and supported by trunnions resting upon girders in the shafts. Hydraulic pumps were furnished for the sump at the bottom of the Lancashire shaft, worked by water from the main pipe, to deal with any leakage that may occur in the tunnel and shafts.*

---

* For a fuller account of the tunnelling operations, *vide* British Assoc. Report, 1892, p. 532.

CHAPTER VII.

THE DISTRIBUTION OF WATER.

360. AMONG the numerous requirements for the efficient distribution of water, are primarily :—Due appropriation of the total quantity supplied among the consumers, according to their needs ; the maintenance of sufficient pressure at the places where the water is delivered, to meet the objects of its use, whether for trade, power, or domestic purposes ; and rapidity of circulation through the pipes of the entire supply-system.

361. The first step, then, in planning the distribution of water, is to ascertain precisely the quantity likely to be, in the future, required by consumers at every point of the district to be supplied. The anticipated future demand at any point exercises an important influence upon the dimensions of and situations assigned to the pipes laid for present requirements in its vicinity. The rate at which the demand for water appears likely to increase, must determine to what extent it is prudent to anticipate the future, by burying pipes of larger size and greater cost than those absolutely required at the present time ; and how far the future may be left to provide for its own needs. (Compare § 321.)

Although it is customary to speak of the average daily supply as a definite quantity of water, which is a highly important figure in respect of the annual provision from the source of supply ; it is necessary, in designing a distribution-system, to consider the variation to which the consumption of water is subject, with a view to provide for the maximum demand whenever it may occur. Such variation naturally depends upon local circumstances, such as those of climate, custom and trade. In the first place, the daily demand during the hottest part of the summer differs, in this country, between 20 and 40 per cent. from the average daily supply. And, in the second place, the draught upon the service-pipes is ordinarily twice as great during the middle portion of the day as the average over 24 hours. These differences are greater, the less the waste of water from leakage and other preventible causes. Further, the fractures of water-pipes and fittings that occur during severe frost, and the exigencies of conflagrations, make it desirable to provide an ample margin of capacity in the service-mains, especially those of manufacturing and of densely-populated districts.

It appears, therefore, that the distribution-system should be planned to be capable of supplying water at a rate equal to at least three times that of the average daily supply of the whole year. The subject of the estimation of the individual demand for water is alluded to in § 392, hereafter. In this country it is not infrequently assumed to be about 25 gallons per head per day (average) for great manufacturing districts, and somewhat less for other places ; but it is well-known that a considerable portion of such a supply as this cannot be used to minister to the personal needs of the consumers.

362. Having determined the maximum rate at which the supply is to be delivered at every point in a given district, the second step is to so adjust the sizes of all the distributing-mains, that this delivery may take place without undue loss of head by friction in the pipes, and other disadvantages contingent upon excessive velocity of the flow through them.

Many towns are situated in localities of such configuration that the head of water necessary to supply their most elevated parts would cause an unnecessary and, in some respects, disadvantageous pressure in districts at lower levels. In such cases, the plan usually adopted is to separate the whole area of supply into zones lying at successively lower levels, carrying independent systems of distributing-pipes branching off the main through each of the zones thus laid out. By this arrangement, the pipes in each zone of supply would experience a higher hydrostatic pressure than those of a superior zone, and little benefit would result from the arrangement, were not means taken to reduce the pressure in the service-main at every step from a higher to a lower zone.

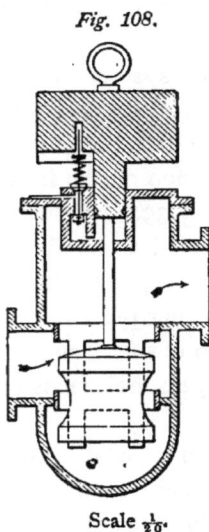

Fig. 108.

Scale $\frac{1}{20}$.

The apparatus employed for this purpose is the "reducing-valve," which is introduced upon the main immediately above the system of distributing-pipes which it is intended to relieve from excessive pressure.

Fig. 108 illustrates diagrammatically a form of the apparatus employed on the Liverpool waterworks for mains of moderate size. It consists essentially of an equilibrium-valve opening downwards, through which the water flows at a rate determined by the opening of the valve and the difference of pressure between its inlet and outlet sides. The valve-spindle passes through a small cylinder and is provided with a piston fitting the latter closely, communication being established between it and the outlet side of the equilibrium-valve by a small hole in the top of the valve-case which forms the bottom of the cylinder. The spindle is carried still higher and is loaded with a weight *W* such that

$$W = A\,p\,;$$

where A is the area of the small piston, and p is the pressure per unit area which it is desired to maintain on the outlet-side of the reducing-valve. If this pressure increases above the assigned value, the piston, and with it the spindle and weight, is raised and the main valve is throttled—the pressure consequently tends to regain its normal condition. If, on the other hand, the pressure diminishes, the piston falls and the valve opens—permitting water to pass more freely and with less frictional diminution of the head or pressure on the inlet-side. In order to throw the valve out of action altogether in the event of fire occurring, when reduction of pressure in the service-main might be a serious draw-back or, at any rate, might well be dispensed with, a second hole leading from the under-side of the small piston is closed by a valve, the spindle of which is struck by the weight of the main valve if the latter descends abnormally.

The position of the spindle of this secondary valve having been adjusted properly, if a fire-hydrant is opened on the outlet side of the main valve, the extraordinary depression thus produced in the latter causes the weight to strike and open the secondary valve—putting the under-side of the small piston into direct communication with the atmosphere, when the unsupported valve drops and opens to its fullest extent. This device is due to Mr. G. F. Deacon.

363. The question of the sizes and strength of pipes has been dealt with in Chapter VI.; and it is only

necessary here to point out that a considerable margin of
strength must be allowed in the case of pipes into which
water is delivered directly from pumps—especially those of
the Cornish type (§ 162).

In calculating the loss of head in distributing- and
service-pipes, the following investigation based upon an
assumption of regular and uniform abstraction of the water
by consumers throughout the entire length of a given pipe,
has been employed by us, and may be found useful :
Suppose that in a certain length of clean pipe, L, water
is drawn off uniformly throughout that length, through a
large number of service-pipes.   If Q is the quantity of
water which enters the length L per second, and P is the
quantity which flows out at the end of that length, then
q, the quantity passing any point, distant l from the beginning, in one second, is

$$Q - (Q - P)\frac{l}{L} \; ;$$

now
$$\frac{dh}{dl} = \frac{q^2}{1850\,d^5} \qquad (\text{§ 72}),$$

therefore
$$\int_{h_1}^{h_2} dh = \int_0^L \frac{q^2}{1850\,d^5}\, dl$$

$$= \int_0^L \frac{dl}{1850\,d^5}\left(Q^2 - 2\,Q\,(Q - P)\frac{l}{L} + (Q - P)^2\frac{l^2}{L^2}\right).$$

Hence, the loss of head, h, throughout the entire length
of the pipe, is

$$\frac{L}{1850\,d^5}\left(Q^2 - Q\,(Q - P) + \frac{(Q - P)^2}{3}\right)$$

$$= \frac{L}{5550\,d^5}\,(Q^2 - P\,Q + P^2).$$

If no water flows out of the end of the pipe, $P = o$, and the loss of head is

$$\frac{L\,Q^2}{5550\,d^5}.$$

A similar investigation applies to rusted pipes, (iv.) § 72.

364. It is often advantageous to lay the primary mains from the service-reservoir or pumps, as the case may be, in duplicate, by which plan the supply is not interrupted during cleansing or repairs of the pipes. The branch (distributing) pipes are then connected with both mains.

Sometimes a duplicate or " rider " main is laid to convey water from the service-reservoir to a distant point, without the loss of pressure which is inseparable from the use of any but very large service-pipes during the busy hours of the day. In extensive districts, duplication of the mains, which might be otherwise desirable, is avoided by the provision of subsidiary service-tanks, situated in elevated positions or upon towers, at suitable points. The main pipes deliver into these tanks, whence the district pipes draw their water according to the demands upon them. The principles that regulate the sizes of service-reservoirs are discussed in § 192. The relative cost will generally determine the particular arrangement to be adopted. Isolated houses situated at high levels, and at the extremities of the distribution-system, have often to rely upon such supply as they can obtain from the service-pipes during the night time, when the draught is generally slight and the pressure consequently high. Where circumstances warrant it, an auxiliary pumping-engine is sometimes erected to raise water to houses so situated from the nearest point to which the service-reservoir delivers it by gravitation in sufficient quantity. By this arrangement the evils of a precarious and intermittent supply of water may be to some extent mitigated.

365. The pipes ordinarily used for the distribution of domestic supply in this country, are cast-iron, of either the lead-jointed or the turned-and-bored type (§ 326). The latter possess the advantage of being readily and cheaply laid ; but should have interspersed among them at regular intervals, not less than 10 per cent. of pipes with lead joints, to allow for movement of the mains due to temperature and to settlement of the ground. They are, in this country, usually laid at least 2 feet 6 inches below the surface of the ground ; and their position is determined according to some definite rule —so that they may be easily found at any time, without the necessity for extensive exploration of the streets or roads under which they pass.

Branch distributing-pipes leading directly off the main, or even secondary pipes when the latter are of large size, are provided with stop-valves at the junctions, in order that when repairs or extensions are necessary, the area from which the water is shut off may be as restricted as possible. In arranging the ramification of service-pipes, it is important, whilst providing means for isolating each of the groups of pipes into which the entire system of distribution is subdivided, to connect one with another in such a manner as to secure free circulation of the water everywhere under normal working-conditions. This is desirable in order to prevent stagnation and consequent deterioration of the water at any point ; as well as to provide a copious supply under the highest possible pressure in the event of fires occurring at a considerable distance from any of the main pipes.

Where the waste-water-meter system (§ 390) is used, the sub-divisions of the distribution-system may be arranged to contain 2000 to 3000 consumers. At the place where the distributing-main for such a district branches off the primary main, a stop-valve is furnished on the former pipe, situated a few yards distant from the branch. A by-pass pipe is conveniently laid to connect branches from the distributing-main on either side of this stop-valve ; and in the circuit of this by-pass the meter is introduced—stop-valves being often inserted on either side of it so that it may be examined internally, or removed, without interfering with the supply.

When the valve on the distributing-main is closed, the entire supply to the district is caused to pass through the meter. In addition to the stop-valve commanding each such sub-division on the main pipe, it is well to control the supply to every 500 or 1000 consumers by a subordinate valve ; so as to be able to still further localise the inconveniences that arise from cessation of supply during repairs and extensions of the water-service.

366. There are three forms of apparatus by which water is ordinarily drawn from the service-pipes for use : Hydrants, street-fountains, and house-service pipes.

The three principal types of hydrant are (1) stopcock hydrants, (2) screw-down loose-valve hydrants, and (3) ball-hydrants.

(1) The stop-cock hydrant, *Fig. 109*, is sufficiently explained by the diagram. It is a convenient form for use as a scour-hydrant at the dead-end of a service-pipe.

(2) The second type of hydrant generally consists of a valve with a single face which is forced against its seat by the pressure of the water. The valve slides loosely on the nut through which the spindle works so as to allow this.

(3) The third class, *Fig. 110*, consists of a vulcanite or indiarubber ball, held up by the pressure of the water in the service-pipe against a leather washer which surrounds an orifice in the top of the pipe. Over this orifice is attached by a bayonet-joint, when it is required to be used, a stand-pipe with connections for fire-hose at its upper end. Down

the centre of the stand-pipe passes a rod, screwed through a fixed guide near the bottom, by means of which the ball may be depressed and the water be drawn from the pipe.

This hydrant possesses the advantages of simplicity

Fig. 109.   *Fig. 110.*

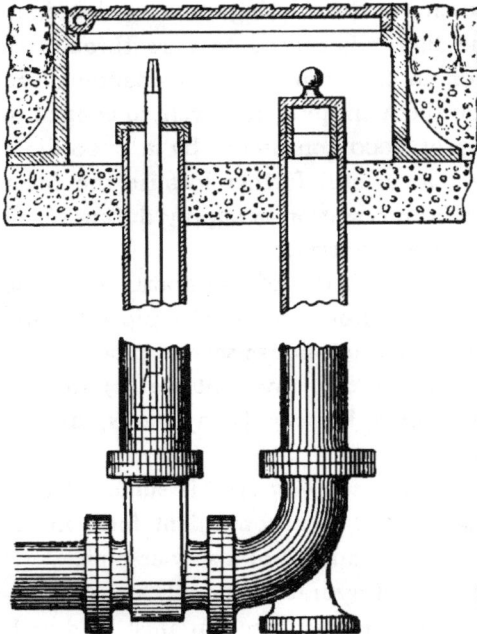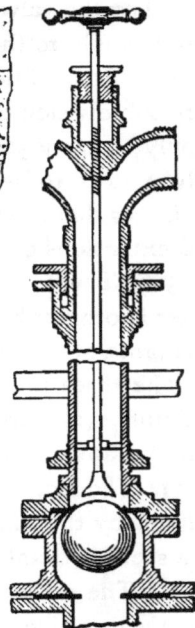

Scale $\frac{1}{12}$.

and cheapness; a draw-back to its use is that the balls are liable to be squeezed out of shape, and leakage is apt to occur in consequence.

In all three types, the screwed connections for the fire- or street-watering-hose are protected by caps when not in use.

For filling watering-carts, a high curved pipe termed a
"swan-neck" is attached to the hydrant, instead of the
ordinary short stand-pipe illustrated in *Fig. 110;* or it may
be attached to a special watering-stand-post protected by
a cast-iron case. The valves of such stand-posts are best
situated below the ground beyond the reach of frost; and
a cock should be provided to completely empty the pipe.

For fire-service, hydrants are placed on the mains at
intervals between 50 yards and 100 yards apart ; but in the
neighbourhood of important buildings and docks they may
be much closer together, and are often put inside houses
on stair-cases or in other con-
venient and accessible situations.

Fig. 111.

367. Street - fountains for
domestic supply, *Fig. 111*, are
provided with self-closing stop-
taps. Otherwise they may lead
to great waste of water. The
closing is effected either by a
quick - threaded screw - down
valve weighted or forced upon
its seat by a spring, or by some
similar simple device. Concussion in the pipes leading to the
fountain may be prevented by
employing a valve with a small
orifice ; and, if the pressure is
high, adding an air-pipe to
that supplying the fountain on
the inlet side of the valve, so

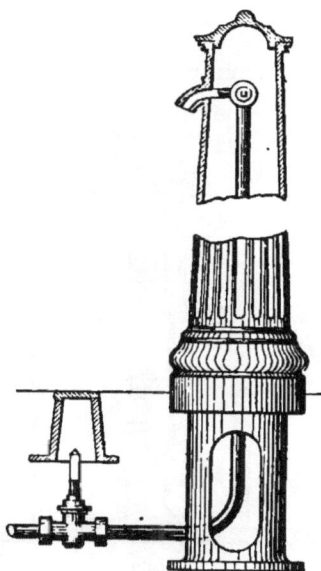

Scale $\frac{1}{10}$.

as to act as an air-vessel in cushioning the ram of the
water when the valve is shut.

368. House-service pipes are usually of lead with wiped

joints, although wrought-iron pipes are sometimes used for introducing the supply through the outer wall of the house.

The sizes and weights of small lead and iron pipes are given in § 331. They are connected with the service-main by brass ferrules screwed into it, *Fig. 112.* The nozzle of the ferrule is furnished with an ordinary union joint, although sometimes this is discarded in favour of a wiped joint.

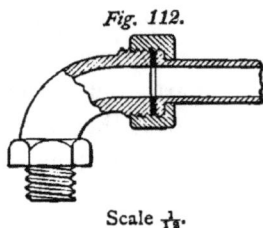

Fig. 112.

Scale 1/12.

369. In order to avoid the necessity for cutting off the supply of water in the service-main whilst ferrules are inserted, the operation is frequently conducted under the full pressure of the main. One machine for this work, *Fig. 113,* consists of a chamber with a horizontal revolving cover, in which are two passages provided with stuffing-boxes equidistant from the centre of revolution. By means of an indiarubber ring, a watertight joint is made between the lower edge of the chamber and the service-pipe that is being operated upon. A combined drill and tap is introduced through one of the stuffing-boxes in the cover and a ferrule through the other. A hole is drilled through the barrel of the pipe, and tapped; the cover is then revolved so as to bring the ferrule over the hole thus made, into which it is screwed.

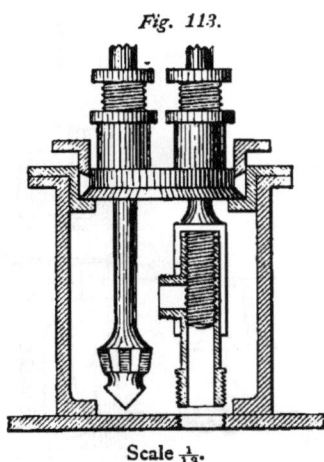

Fig. 113.

Scale 1/12.

The chamber and the rest of the apparatus can then be removed. The ferrule employed for this purpose, has, in addition to the ordinary right-angled branch, a vertical hole throughout its length, into which, before the ferrule is inserted, a plug is screwed from the top so far down as to close the branch. When the house-connection has been completed, this plug is withdrawn as far as is necessary to permit free passage of the water through the ferrule.

Another method of effecting this operation, due to Mr. Oswald Brown, consists in first drilling and tapping a

Fig. 114.

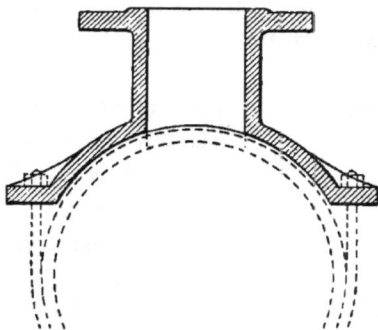

hole nearly through the service-pipe, and screwing in a ferrule of similar form to that last-described except that the branch instead of the top is plugged. A drill is inserted through a temporary stuffing-box on the ferrule, the hole in the pipe-barrel is completed, the drill is withdrawn nearly to the top, and the side-plug is screwed up. The drill may then be re-placed by a long screw plug, the plug in the branch being taken out and the house-connection completed as in the method previously described.

370. When a service-pipe of too great size to admit the employment of a ferrule, or a branch-pipe has to be

connected with a main of such dimensions that it is practicable to cut a large enough hole into it without injury, a saddle-branch, *Fig. 114*, is fixed upon the main.

This consists of a short segment of pipe with a flanged or socketed branch of the required size and shape. The hole is cut in the position required. A flat indiarubber or gutta-percha ring is placed between the saddle and the main pipe, and the former is fixed either by studs through the saddle tapped into the pipe, or by two wrought-iron straps terminating in screws, which pass round the pipe and through holes in flanges formed upon the lower edges of the saddle, *Fig.* 114.

371. At every junction of a house- or trade-service-pipe with the main, a stop-cock, *Fig. 115,* is fixed; and its position with respect to some permanent object such as a painted mark upon a neighbouring wall is carefully measured and recorded. This precaution should be observed, too, in the case of every hydrant and valve in the distribution system. An indication of the positions occupied is frequently given by a painted figure at each point, and is most useful.

Fig. 115.

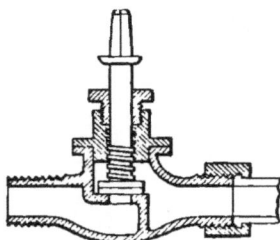

Scale $\frac{1}{15}$.

Cast-iron surface-boxes with hinged and locked lids are used to cover, whilst rendering accessible, hydrants, valves and stop-cocks. They are supported quite independently of the apparatus they cover, in order that the latter may not be affected by any superincumbent weight. They are made as small as is consistent with facility for putting the key on a valve or attaching the stand-pipe to a hydrant. Their shape is such as to accommodate the packing of road-metal, of whatever kind may be in use around them.

372. For domestic supply, a pipe ⅜ inch to ¾ inch in diameter is sufficient, except for houses of the largest class. The pipe is led into the house and up to a cistern, covered to exclude light and dust and placed in an accessible and cool place protected from frost. A cistern for use under constant supply, by which is meant that the supply is always available, at full pressure, to the consumers, is situated above the level of the hot-water and cold-water taps : at least one of the latter and any other cistern-supplies are connected with the feed-pipe to the cistern. In intermittent supply (§ 376 *post*) also at least one tap should be connected with the feed-pipe on its way up, in order that water for dietetic purposes may be obtained without passing through the cistern.

The capacity of the latter must be to a certain extent dependent upon the character of the water-supply; but in no case ought it to be necessary to store more than 24 hours' supply in any dwelling house supplied from a public source.

The cisterns are often made of wood lined with sheet-lead weighing 5 lbs. per square foot ; but iron or slate cisterns are far superior. Each one should be provided with a ball-tap not submerged when it has completely shut off the water. The level of the water in the cistern should then be 3 inches below the overflow-level. The overflow-pipe should be brought to a conspicuous place outside the building, to act as a warning pipe ; but it must not dis-charge over any ashpit or near any other source of pollution.

For hot-water supply, a pipe is taken from the bottom of a small auxiliary cold-water cistern, provided sometimes with a check-valve to prevent hot water from rising into the latter.

In intermittent supply, the various cold water taps and flush-cisterns would be connected to this pipe. It is introduced into the hot-water cistern at the bottom. The hot-water pipe is taken out of the top of the hot-water cistern, and supplies all hot-water pipes on its way upwards. Its open end is bent over the top of the auxiliary cold-water cistern, so that it is impossible for a greater pressure to obtain in the hot-water pipes than that due to their height of water, and damage from overflow is prevented. The size of the hot-water cistern depends upon the quantity required at one time. To heat the water in ordinary houses, a small wrought-iron or copper boiler is generally fixed at the back of the kitchen range. A pipe from the top of this boiler enters the hot-water cistern near the top, and a pipe from the bottom of the latter enters the boiler near the bottom. A constant circulation of water is maintained between the two when the boiler is heated.

373. A flushing-cistern for a water-closet or urinal, should be fixed at least 4 feet 6 inches above the latter with a $1\frac{1}{4}$ inch down-pipe. The cistern should be capable of giving a flush of 2 gallons in 15 seconds through the down-pipe into the basin. It may be of either the double-valve or the siphon type.

The former consists of two chambers connected together at the bottom by a passage which is closed in one chamber by a weighted valve with an india-rubber or leather washer; in the other chamber the weighted valve closes the outlet to the down-pipe. The water flows into the first chamber through a ball-cock and fills the second chamber, the supply being shut off by the ball-cock, when both are full of water. The spindles from the valves are furnished with slots in which work pins projecting from arms attached to the same axis. These are so arranged that

it is impossible for both valves to be either on or off
their seats simultaneously. When not in use, the valve in
the first chamber is off its seat, whilst that in the second
chamber closes the flush-pipe. In action, the valve in the
first chamber is depressed so as to cover the opening
connecting the two chambers, whilst the valve in the
second chamber is simultaneously lifted off the opening to
the flush-pipe and that chamber discharges its contents.
The reverse action of the valves is then automatically
effected by counter-poise, the water runs from the first into
the second chamber, and the ball-cock opens and again fills
both of them.

The siphon-cistern, *Fig. 116*, consists of a single

Fig. 116.

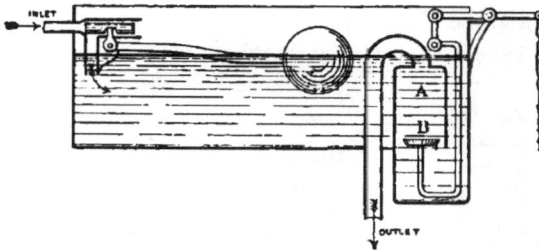

Scale ¹⁄₁₂.

chamber, in 'which the water-level is regulated by a ball-
cock. It should be capable of holding 2 to 3 gallons of
water. The down-pipe through which the flush is delivered
is bent into the cistern below the water-line, and then
widens out to form a cylindrical chamber, in diameter
about four times that of the pipe. This chamber, at the
foot of which is a piston capable of moving upwards,
reaches to the bottom of the cistern. When the piston
rests on its seat, water can enter the chamber through an
orifice immediately above the piston. When it is lifted, the

upper bend of the down-pipe is filled by water from the
chamber and the cistern is siphoned empty. Of course,
as soon as the water-level begins to fall, the ball-cock that
supplies the cistern opens; but the quantity of water which
enters the latter during the 15 seconds occupied in emptying it is very small.

A ball-valve generally consists of a nozzle at the end
of a supply-pipe, closed by an india-rubber or leather
washer mounted in a metallic plug, which slides backwards
and forwards in a cylindrical guide projecting from the
nozzle. The plug is actuated by a lever from a copper
ball-float, which forces the plug against the nozzle as it is
borne up by the rising water, and stops the flow at 3 inches
below the overflow-level. Ball-valves for household-supply
are $\frac{1}{4}$ inch to $\frac{1}{2}$ inch in diameter, according to the pressure,
and are made of hard brass.

374. When the supply is taken from a house-cistern,
plug-cocks may be used; but when-

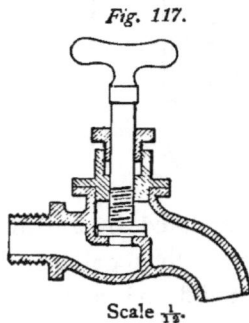

Fig. 117.

Scale $\frac{1}{13}$.

ever water is drawn direct from
the service-pipes, screw-down bib-
cocks, made of brass or gun-metal
with gun-metal spindles, must be
used (*Fig. 117*). The valve is
sometimes attached to a small
spindle which works loosely in a
socket in the screw, so that the
water-pressure holds the valve up
when the cock is open. When the
screw is turned to shut down the valve, the leather washer
is pressed home, but is not rotated after reaching its seat.
The valves of cocks buried in the ground work loosely
on the spindles, but cannot become detached from them;
so that when the upper portion of the cock is lifted out
for repairs, the valve does not fall away. Instead of

leather, rotating metal washers are employed with good results, as in the tap invented by Lord Kelvin.

375. Hitherto, we have tacitly assumed the existence of two main conditions of supply—the maintenance of constant pressure in the service-pipes with unrestricted access by the consumers to the water therein, and free use of the water for all domestic purposes, without the restriction indirectly exercised by a system of payment according to the quantity used.

In some towns in America, the colonies and on the Continent, water is supplied to the consumers entirely by measure.

Paris, Vienna and Berlin are examples of cities thus supplied. In the last mentioned the whole supply is delivered through the Siemens' meters (§ 109). In Berlin and Vienna the houses are built in flats, and average 67 persons per house, each supplied by a separate service-pipe and meter. There is an absence of house-cisterns in Berlin, and the city is otherwise peculiarly adapted to receive a meter-supply.

It is to the general advantage of the whole body of water-consumers in any given district, that the service be efficient in every portion of it. Waste of water, and the concussion caused in the pipes by the rapid closing of taps, are not only adverse to the interests of the individuals at fault and to the Water Authority, but affect also in some degree all the consumers of the district in which such mischief occurs. Hence an important duty of a Water Authority is to exercise due supervision over all the apparatus and fittings used in connection with a public domestic water-supply. None should be allowed to be employed, except such as are structurally efficient and have been found capable of sustaining proper proof-tests.

With regard to the soundness of the plumbing, and

other work, in connection with the water-service, a great deal may be done by the registration of all plumbers who agree to conform to the waterworks regulations, in return for the privilege of having their names advertised on the printed forms of regulations, and in some of the official notifications to householders. Such regulations stipulate that all new fittings shall before use be tested and stamped by an official testing-officer, and that new work, before being covered up, must be examined by an authorised representative of the Water Authority.

376. When no efficient steps are taken to prevent waste from defective fittings and pipes, an "intermittent" system of supply is sometimes adopted—the water being shut off the houses during a large portion of the day. This prevents the loss from leakage during that time, though at the expense of rendering the water unavailable for legitimate use, and in case of fire. Now that methods of detecting and preventing waste are so well understood, this plan of reducing it should never be resorted to. In fact no cause other than unavoidable deficiency in the supply, is a sufficient reason for adopting it.

The intermittent system of supply necessitates the introduction of large cisterns (§ 267) which are seldom kept properly clean; and in the poorer class of houses, they cannot often be so placed as to be easily accessible and kept cool, whilst they are often subject to the influences of an impure atmosphere. In the poorest districts, where the water is obtained from fountains, it has to be drawn and left standing for many hours in crowded rooms. These sources of pollution, together with the increased temperature, however unimpeachable the quality of the water in the distributing mains, cause its quality to be in many cases much deteriorated before it is used. When the water is turned off, under such conditions, the pipes partially empty through leaks and fractures at the lower levels ; whilst sewer-gases or objectionable liquids are often sucked in at other parts of the system—causing the water, when the pipes are re-charged, to be tainted.

377. It is, in sea-coast towns, often found convenient and economical to provide a low-pressure supplementary supply of sea-water for road-watering, sewer-flushing, bathing and other purposes. If circumstances admit of it, an effective and cheap fire-service may also be thus furnished. In districts situated near to great rivers, the waters of which are deemed to be unwholesome for dietetic use, and so far removed from sources of pure supply as to render the cost of the latter considerable, the question of duplicate supply becomes important. It is well known that the personal consumption of water for culinary and drinking purposes, is a small fraction of the total demand ; and it would frequently appear to be an economical course to procure merely the needful supply of the purest water for this service, and to furnish a secondary supply of cheaper water for washing, trade and public purposes. Apart, however, from the difficulty and intricacy necessarily involved in such a dual system of water-supply, the question must always arise in such a case—will the classes who are most benefited by an excellent supply of water, use the necessary discretion and care in discriminating between the proper and the improper employment of the two supplies ?

Where it is available, the most satisfactory system is, undoubtedly, to supply only water of unexceptionable quality, and to economise by the avoidance of its misuse and waste.

2 D

CHAPTER VIII.

THE MAINTENANCE OF WATERWORKS.

378 EVERY scheme of water-supply, whether for industrial purposes or to meet domestic wants, involves, at the moment of its completion, the question—how is the supply procured to be satisfactorily maintained? Although this subject is, as a matter of convenience, treated after that of the construction of the works; it forms, in effect, a vital element of the initial consideration bestowed upon every such undertaking. The capitalised cost of the outlay incurred in maintaining the efficient operation of the system of supply which is to be adopted, and in the necessary repairs to the works, must be duly considered at the outset; and, in a comparison of several rival schemes, the prime cost of construction of the works must be regarded as only one of the factors that are to determine their relative merits. Experience of both the actual construction and the maintenance of waterworks, can alone afford the data necessary to the formation of a sound judgment hereupon, or warrant an engineer in exercising preference in favour of any particular project.

379. No matter how excellently waterworks may have been originally built, the unceasing operations of atmospheric agents, and the wear and tear of machinery, demand continual renovation of the works. The more continuous such repairs are, the slighter is their character. Allusion in detail to the subject of renewals and repairs,

could scarcely amount to more than a recapitulation of the methods of construction already described, and an essay upon numerous items of mechanical arts that are outside the scope of this treatise.

All appliances such as sluices, valves, hydrants, air-cocks, street-fountains, meters, &c., must be subject to constant inspection by qualified mechanics, and any derangement in them repaired as soon as discovered ; otherwise the mechanism, exposed to the unfavourable conditions that obtain in ordinary waterworks practice, will not long continue to perform the duties assigned to it. All valves should be occasionally worked through a portion of their stroke, in order to prevent them from sticking fast.

The care and attention demanded by pumping-machinery of all kinds does not need enforcement here.

Exposed metal-work must be scraped and painted regularly. It is of importance that a sufficient stock of both ordinary and special pipes and of other apparatus be kept at convenient places along the line of a supply-main as well as in the districts supplied ; in order that there may be as little loss of time as possible in effecting repairs when breakages occur. On all long lines of aqueduct, it is usual to provide for the daily inspection of the pipe-line by " walks-men," whose business it is to perambulate the several sections of the route assigned to them, to attend to the valves, and to give immediate notice to the engineer of failure in or accident to the portion of the works under their charge. Fractured pipes when found are cut out (§ 330), or their joints, if of lead, are melted out by fire—a method that is, however, not easily applicable to large mains. New pipes, when inserted, are linked in with the old work by collar-joints (§ 341).

The periodical cleansing of reservoirs, tanks and cisterns, and the frequent flushing of pipes by scour-valves and hydrants, are operations that need no special commendation to those who appreciate the highest possible standard of purity in water-supply.

380. There is one cleansing-process applied to water-mains which calls for special notice. The corrosion of iron pipes and the measures taken to prevent, or rather to delay the commencement of, this action, have been already alluded to (§ 328). Such corrosion arises from the action of water upon the metal ; and, as might be expected, is found to proceed more rapidly in pipes in which there is a continuous and rapid flow than in those wherein the water is comparatively stagnant. In pipes that are not protected by a proper preservative coating, corrosion begins at once to affect the entire surface, and attacks the metal vigorously ; but in well-protected pipes, the action may be long deferred—the most minute bare places, however, affording situations for the development of nodules of rust, which, if numerous, gradually burst off the intermediate coating and coalesce. The rust thus formed consists principally of ferric oxide associated with considerable amounts of organic and of siliceous matter—the ferric oxide being produced by the formation of ferrous carbonate and its subsequent solution in carbonic acid. Hence it is, that soft water which naturally carries much free carbonic acid derived from the atmosphere, is peculiarly destructive to pipes and other apparatus made of iron.

This internal corrosion is always a source of some trouble, and if not anticipated at the outset and properly provided against, leads to "furring" of the pipes, with the concomitant inconveniences of reduced supply of water and diminished pressure in the mains, and sometimes to the expense of early renewing this portion of the works.

381. Rust may be detached from the metal by the application of heat ; but, in practice, the course generally

adopted for the cleansing of corroded mains is one of the following :—

(1) The pipe is severed at intervals, and the rust is cut out of it by a tool of the shape shown in *Fig. 118.*

This consists of a steel bar 3 feet or 4 feet in length with a cutting point at the turned-up end, which is forced

Fig. 118.

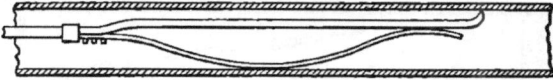

Scale $\frac{1}{20}$.

against the pipe-barrel by a spring riveted to it at one end and bearing against the opposite side of the barrel. The handle is formed of sections of tube of convenient length, screwed or otherwise strongly coupled together.

This cutter is worked forwards and backwards by hand, being gradually turned so as to traverse the whole internal surface of the pipe. A flush of water is employed to wash the detached rust and scale down to the end of the length of pipe operated upon ; and it is then raised out of the pipe-trench by a chain- or a bucket-pump (§ 156). The cleansing of bends by this method presents much difficulty ; and it is only applicable to small pipes.

(2) By another process, the detachment of the rust is effected automatically by a tool driven along the main by water-pressure. A common form of the apparatus employed for the purpose is illustrated in *Fig. 119.* It consists of four or more barbed steel cutters, carried by stiff springs which are bolted to an iron bar about 3 feet or 4 feet in length. One end of the bar is attached by a universal joint to a piston or disk of diameter somewhat less than that of the pipe to be operated upon. The other end is similarly attached to a ring of the same diameter.

The piston is faced with a leather or india-rubber washer to make a joint with the pipe-barrel. Rings for the attachment of cords or chains, to assist or control the motion of the machine, as is frequently found necessary in dealing with foul pipes of less diameter than 9 inches, or to effect its withdrawal, are provided at each end.

Fig. 119.

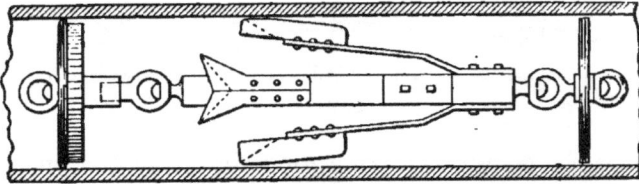

Scale $\frac{1}{20}$.

When used, it is introduced into any given section of the pipe through a man-hole or hatch-way, and the cover is bolted on. The stop-valve at the distant end of the section in question is shut, the main is charged, and one of the lower intermediate scour-valves is opened. The pressure upon the piston of the machine moves it along, the cutters scaling off the rust, which is washed down the pipe and discharged through the scour-valve. The rate of motion is governed by the manipulation of the latter. It is generally advisable to pass the tool twice through each section of the main which is treated. Other useful forms of this apparatus are often employed, in some of which a helical motion of the cutters is obtained with a view to accomplish the cleansing as uniformly as possible and by a single operation. Sharp bends naturally introduce difficulties in the application of this process, and attempts to pass round them are liable to result in jamming the tool fast, and the consequent trouble of cutting it out of the main.

An interesting account of some work with these machines, by Mr. H. P. Boulnois, may be referred to.*

382. So far, we have alluded only to the comparatively simple operations of repairing and maintaining in good condition the apparatus comprised in a waterworks system. But, in the management of every such undertaking, by far the most difficult task that falls to the lot of the engineer, is that of detecting existing imperfections in works either hidden underground or otherwise not easily accessible to inspection, and of preventing their continuance. The history of many waterworks undertakings may be summarised thus : Ample provision for the needs of the consumers at the outset, without limitation of the quantity used by any individual ; gradually increased consumption due to the more lavish use of water, the growth of population and the expansion of trade, and the development of unseen leakage from the mains and service-pipes ; restriction by the Authority, of the hours of supply—at first exercised as a means of husbanding its resources, but subsequently used as the only alternative to immediate famine ; inconvenience to trade and injury to the health of the consumers ; finally, rigorous suppression of the waste and misuse of water, and the inception of new works to augment the supply to meet the necessities of the case.

383. When the consumption of water outgrows the available supply, there can be little question as to the propriety and prudence, as there certainly is no doubt about the effectiveness, of economising water by simply reducing the number of hours during which it is supplied. This plan, however, involves in the first place the somewhat illogical course of treating alike the legitimate use, misuse and leakage of water ; and, whilst suppressing somewhat the evidence of the evil, it does nothing to

* Report on Cleansing the Rising-mains of Exeter—Exeter, 1882.

remedy it.   Beyond this, it has an objectionable result that
has been already alluded to (§ 376), but is so important as
to warrant some reiteration.

384. During the hours when the water is cut off, the
distributing-mains are emptied by the consumers ; and
the partial vacuum thus caused in them cannot fail to suck
in through fractures and imperfect joints a daily charge of
subsoil impurities, which, at any rate, have little chance of
entering the most leaky water-pipe if it is under constant
high pressure.

Moreover, it must be borne in mind that the poorer
classes of consumers have seldom proper utensils available
for storing water drawn from the mains during the hours of
intermittent supply, to be used for their domestic needs
during the remainder of the 24 hours ; and all classes suffer
in various degrees from the use of water stored in dwelling-
houses at high temperatures.   The hardships and insanitary
conditions which result from intermittent domestic water-
service have been so often pointed out, that we will not
labour to re-prove them here ; but will premise that
constant service is only to be departed from when circumstances compel the adoption of such a course.

385. In order, then, to control the proper use of water
and to prevent leakage from pipes, taps and cisterns, the
engineer must adopt and pursue some definite method of
enquiry, to ascertain where waste of water exists.   And,
having localised such waste, he must, if it occurs upon
private premises, be provided with special powers for
dealing with the matter in a somewhat more expeditious
manner than he would by ordinary law be warranted in.

Parliamentary powers are almost invariably bestowed
upon Waterworks Authorities, to authorise them to take
proper measures for the prevention and the suppression of
misuse and waste of water, and penalties may generally be

exacted from persons who offend in this respect. By far the simplest and most expeditious process employed by Water Authorities for enforcing the observance of their regulations upon recalcitrant consumers, is, however, to cut off the supply of water from such persons ; and the legality of this course is generally recognised.*

386. As a matter of public policy, it is, nevertheless, a questionable proceeding. The persons whose neglect or lawlessness leads to open rupture of their relations with the Water Authority are generally precisely those whose indifference to the use of water, as well as to all other sanitary observances, constitutes a standing menace to the health of the community in which they reside. It is surely undesirable to restrict the ability of such persons to use water for legitimate purposes, as a consequence of the means taken to prevent them from using it improperly.

387. Defective house-fittings and carelessness on the part of consumers are by no means the only causes of waste of water. Sometimes considerable leaks, generally indicated by local reduction of pressure, occur in the service-pipes. If the latter are laid in gravel, the water may readily escape through it unnoticed, and find its way into some of the deeper sewers. If the pipes are laid in clay, any escape of water generally rises directly to the surface, that being the line of least resistance ; but, in the case of pipes under streets, covered by strong impervious pavement, this increased resistance is apt to cause the water to find a passage into the basements of adjacent buildings or into the sewers. In order to ascertain the position of such leaks, holes may be bored in the ground with an auger, and the relative levels at which the water stands in the holes noted. The hole in which the water rises highest is nearest to the leak. By boring more holes in the

* Waterworks Clauses Consolidation Acts, 1847 and 1863.

direction indicated, the position of the leak may be approximately determined.

388. The most direct method of ascertaining the existence of domestic waste of water is to frequently inspect the consumers' pipes and fittings. This proceeding is, however, not only very costly and highly objectionable to the majority of householders; but its efficiency depends upon the zeal, honesty and other personal qualifications of the inspector. Beyond this, however, it affords little or no indication of the waste resulting from underground or other hidden leaks in the distributing-mains and service-pipes.

389. A better plan, which early commended itself to those who paid attention to this subject, is to ascertain by an integrating water-meter, the total quantities of water consumed in some prescribed period by several divisions of a district under supply. Thence, by comparison of the figures with respect to the population of the several divisions, the location of centres of waste may be inferred.

By such a method applied at Glasgow in 1872, imperfections of pipes and fittings were disclosed, the repair of which, when effected, reduced the consumption of water in three experimental districts, from 60 gallons, 45 gallons, and 77 gallons per head per day, to 38 gallons, 20 gallons, and 50 gallons per head per day, respectively.*

390. The further development of this plan, and the invention of a meter that furnishes a continuous record of the rate at which water passes through it, has resulted in a most elegant system of detecting waste; which has not failed, when properly applied, to conduce to the efficient and economical maintenance of waterworks in every part of the world.

The originator of what is known as the waste-water-meter system is Mr. G. F. Deacon, who first introduced it

* Minutes of Proceedings Inst. C.E., vol. lxvi. p. 344.

at Liverpool in 1873–75. The meter has been already described (§ 111), and we are now only concerned with its application for the detection of waste.

It is assumed that the reticulation of the mains and service-pipes belonging to a given waterworks, and the disposition of the stop-valves, are in general accordance with the outline of a good distribution system, as sketched in § 365. Any desired district then may, by suitably manipulating the valves that command it, be so far isolated from all others around it, that the supply of water to its service-pipes is derived entirely from one main. The supply is caused at that place to pass through a waste-water-meter fixed on a by-pass pipe in the manner explained in § 365.

If, under these circumstances, the meter-diagrams indicate, during some portion of the night or the small hours of the morning, that a considerable steady consumption of water appears to be taking place in a district where there is no apparent reason for it at such a time— suspicion is at once aroused. It may be reasonably conjectured that such apparent consumption of water in, for example, an ordinary residential district, must be largely due to leakage from distributing- or service-pipes or to waste from fittings such as taps and cisterns.

The diagram illustrating such a condition of things is of the type shown by the fine line in *Fig. 120,** which indicates the consumption (at the rate of 54 gallons per head per day) of water during 24 hours in a small Midland town of 2800 inhabitants on a first application of the method of waste-detection described. The waste-inspectors are thereupon instructed to examine the district for waste. The examination takes place at night. The inspectors first sound in turn the outside stop-cocks of all the premises supplied through the meter, using the valve-keys as

---

* Furnished by the courtesy of Mr. William Hope, of Liverpool.

Fig. 120.

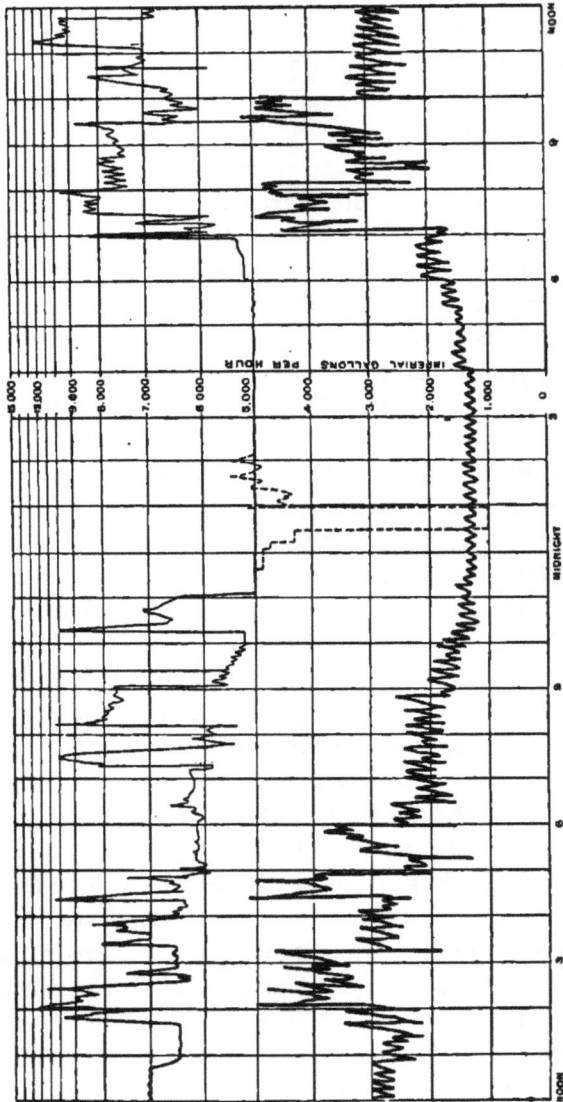

DEACON WASTE-WATER-METER. DIAGRAM, ONE-FOURTH SIZE, ILLUSTRATING WASTE OF WATER, APPLICATION OF THE TEST, AND REDUCED CONSUMPTION AFTER REPAIRS TO DISTRIBUTING-PIPES.

stethoscopes. Every stop-cock through which water is heard to pass, is shut down and marked, and the time is noted. If, after closing the stop-cocks, the sound continues to be heard, it may be inferred that leakage is taking place in the distributing-mains, and the ground is carefully sounded along the line of these pipes until a place is found where the noise attains a maximum. An exploration of the pipes at that place on the following day will probably disclose the existence of a leak arising from a fractured pipe or ferrule. The main stop-valve commanding the entire district is next closed for a few minutes only, the zero-line of the diagram being thereby verified and the condition of the distributing-mains between the house-service stop-cocks and the meter being shown. Finally, all the previously-closed stop-cocks are re-opened, to restore the service throughout the district.

On the following day, the chief waste-inspector examines, by comparison with the indications of the meter-diagram before him, the extent and accuracy of the work that has been done, as recorded in the night-inspectors' report-books. Thence he is able to ascertain the locality of the several leaks that may have been found to exist, and to form an idea of their extent. Instructions are then issued for the visitation of those premises only upon which waste has been proved, and for the necessary repairs to be made to pipes and fittings. By such means "the work of many days' inefficient house-to-house inspection is efficiently performed in one."* For purposes of comparison, a normal meter-diagram from the same district as that to which the fine line, *Fig. 120*, refers, taken some time after the application of the system and consequent repairs, is superposed upon the diagram, and is shown by the heavy line in the same *Fig.*

* Journal of the Society of Arts, 19th May, 1882.

This latter line may serve the further instructive purpose of showing the normal diurnal variation in the rate of
consumption of water in such districts as that referred to.
The supply in this case works out to 18 gallons per head
per day.

391. Among methods of reducing waste otherwise than
by limiting the hours of water-supply, that of regulating
the pressure in each district by means of the reducing-
valve (§ 362) may be noticed. This is, of course, far
inferior to the plan described in the foregoing section ;
because it merely mitigates the action of the evil, without
attacking its origin.

392. The subject of the use and misuse of water, apart
from that of waste arising from defective pipes, cisterns
and taps, is intimately connected with the climatic conditions and geological formation of any given locality, and
with the habits of the residents. Hence, on this account,
analogies between the requirements of widely separated
districts are frequently imperfect, and each one must be
studied on its peculiar merits. The demands of trade and
of shipping are variable but highly important factors in
fixing the proper amount of a water-supply ; whilst, owing
to the difficulty of assessing with accuracy the population
of a given district at any stated time, and to the want of
uniformity in the method adopted to measure the water
supplied by the various waterworks, such statistics as are
available can only be regarded as approximately correct,
and cannot be made a basis for exact calculation.

393. Whilst, however, discountenancing hasty generali-
sation from mere statistical data, as to the requisite amount
of water to be provided in any given case ; we think it may
be instructive to present here a summary of the particulars
of the water-supply of a few towns in Great Britain—pre-

senting a considerable range in point of size and industrial and commercial character.

STATEMENT OF WATER-SUPPLY OF CERTAIN TOWNS IN GREAT BRITAIN.*

Town.	Estimated Population supplied.	Supply in Imperial gallons per head per day.	Nature of the Source of Supply.
London	5,240,000	33	Rivers, springs and wells.
Manchester	950,000	22	Catchment.
Glasgow	850,000	50	,,
Liverpool	770,000	26	Catchment and wells.
Birmingham	640,000	22	Rivers and wells.
Bradford	400,000	25	Catchment.
Edinburgh	350,000	40	,,
Nottingham	240,000	18	Wells.
Brighton	180,000	33	,,
St. Helen's	70,000	20	Wells.
Bath	60,000	20	Springs.
Torquay	45,000	30	Springs.
Chester	40,000	28	River.
Exeter	40,000	38	,,
Scarborough	30,000	23	Springs and wells.
Leamington	25,000	19	Wells.

394. For the sake of contrasting with the foregoing supplies, those of other waterworks systems differing considerably from them, we append similar information with regard to a few principal towns in the United States, and some towns of less size situated in the State of Massachusetts. The latter are selected because they are well

* Report of Roy. Commission (1892) on the Water-supply of the Metropolis —London, 1893. Hastings, 'Waterworks Statistics'—London, 1892. De Rance, 'The Water-supply of England and Wales'—London, 1882.

authenticated, show a considerable range in size, and belong to a State that has exhibited a notable solicitude for sanitary progress.

STATEMENT OF WATER-SUPPLY OF CERTAIN TOWNS IN THE UNITED STATES.*

Town.	Estimated Population supplied.	Supply in Imperial gallons per head per day.†	Nature of the Source of Supply
New York	1,700,000	55	Rivers.
Philadelphia	1,000,000	80	,,
Chicago	875,000	110	Lake.
Boston, Cochituate Works ..	380,000	65	,,
,, Mystic Works	110,000	60	,,
St. Louis	400,000	64	River.
Washington	160,000	125	,,
Worcester, Mass.	69,000	56	Catchment.
Lowell ,, ...	65,000	63	River.
Cambridge ,, ...	60,000	55	Lake and River.
Fall River ,, ...	57,000	26	Lake.
Lynn ,, ...	46,000	45	Catchment.
Lawrence ,, ...	39,000	57	River.
Springfield ,, ...	38,000	88	Catchment.
Holyoke ,, ...	28,000	65	,,

395. A question that affects in an important degree the maintenance and management of waterworks, is that of the sale of the water to the consumers.

For irrigation, power and other industrial purposes, it appears to be generally conceded that a sale of water by measure is the most just and convenient way of gathering

* 'Reports on the Water-supply of the United States'—Washington, 1887. 'Examinations of the Water-Supplies and Inland Waters of Massachusetts' —Boston, 1890. 'Manual of American Waterworks'—New York, 1890.

† It must be borne in mind that the U.S. gallon, which is the unit employed in American statistics, is equivalent to 0·830 Imperial gallon.

the revenue from the consumers. But in the case of domestic supply in this country, it is otherwise.

Even where waterworks are not in the hands of companies seeking to make profits upon money invested in the undertakings, it is considered indispensable that the revenue derived from the consumers for their use of the water should meet the daily working-expenses, the interest upon the capital outlay, and the requirements of a depreciation or redemption fund.

Apart from questions as to the absolute propriety of various kinds of assessment—which not unnaturally accord with the views of individuals, in proportion to the relief they appear capable of affording to individual rating—there is the debatable point, whether the revenue should be derived by a method of taxation at all, or should be obtained from a sale of the water by measure. The legislature of this country has shown itself consistently opposed to the latter method, in so far as it affects the domestic consumption of water.

On the one hand, it is probably impossible to contrive any system of taxation whatever, no matter how ingenious it be, the incidence of which shall fall with perfect equity, day by day, upon the same individuals. On the other hand, the difficulties inseparable from attempts to measure domestic water-supplies with extreme accuracy have not been entirely overcome. Further, if the approximation to the truth given by some water-meters were deemed satisfactory, the cost of introducing such a system into towns in this country would be often prohibitive.

396. Assuming the staff-charges in a town supplied by water-meters to be no heavier than those in a rated town of equal size—a supposition that probably errs in favour of the former system—consider the cost of the introduction and maintenance of the apparatus employed : A reliable

water-meter suitable for an ordinary house can hardly be procured and fixed for less than 3*l.* ; and the cost of keeping it in good order, or the charge for depreciation, may be safely taken at 4*s.* per annum—which, capitalised at 4 per cent. per annum is 5*l.* In order, therefore, to apply the system of supply by measure, the capital expenditure on any given waterworks would have to be burdened with a charge of 8*l.* per-house, in respect of the mere collection of the revenue.

Beyond this, there is a strong feeling in this country in favour of encouraging the unstinted use of pure water by all classes of the community ; and that practice would probably suffer a considerable check, if water for domestic use were sold by measure.

397. Like that of many other undertakings, the successful management of waterworks depends largely upon the personal qualifications of the administrative officers. But, although good administration of laws that are of even indifferent quality, may result satisfactorily; there can be no doubt that the intelligent application of a well-devised code may, in such special and artificial circumstances as we are now concerned with, result more favourably still. One of the last duties of the executive engineer is generally to frame regulations for the future maintenance of the works he has constructed.

This may, therefore, be deemed an appropriate place to introduce an outline of the general provisions of waterworks regulations that have been tested by prolonged practice and are found to answer well:

(1) Every house or tenement supplied shall be provided with a separate service-pipe laid at a depth not less than $2\frac{1}{2}$ feet below the surface of the ground.

(2) Each service-pipe shall be, unless otherwise especially agreed upon, made of lead (of the weights stated, in an

appendix to the regulations, to be appropriate to its size and the pressure that it will be subject to). It shall be attached to the service-main in the street by a brass ferrule screwed into the iron pipe and fixed by the Water Authority. The service-pipe shall be furnished with a stop-cock set in a cast-iron box situated outside the premises supplied.

(3) The joints of the lead service-pipe shall be of the kind known as "wiped" joints.

(4) A cistern of a capacity of not less than 25 gallons must be provided for every domestic supply. Such cisterns should be of iron or slate, although wood lined with sheet-lead weighing not less than 5 lbs. per square foot is some-times permitted.

(5) Each cistern shall be provided with a ball-tap that has been duly tested and proved water-tight under a pres-sure of 300 lbs. per square inch ; with an overflow discharg-ing through the wall of the house in such a position as to be easily observed without ; and with a ground-brass valve at the bottom, discharging in a similar conspicuous position.

(6) Boilers, water-closets and urinals shall be in every case supplied from separate cisterns. Cisterns attached to water-closets and urinals shall be capable of discharging 2 gallons of water in each complete flushing-operation in 15 seconds.

Experience has shown that a most fruitful source of the misuse of water lies in defective flushing-apparatus. It is essential that flushing-tanks or -cisterns be con-structed of the dimensions appropriate to the objects they are to serve ; and that they be so arranged as to ensure the rapid discharge of their entire contents without further draught upon the source that supplies them, until that operation has been completed. Flushing-apparatus of this kind is described in § 373.

2 E 2

(7) Every bath shall be furnished with a water-tight ground outlet-plug, firmly attached to it by a chain.

(8) Every street-fountain or stand-post shall be provided with self-closing apparatus incapable of being shut suddenly.

(9) All taps shall be of the screw-down type, incapable of being closed suddenly.

(10) Air-vessels of approved dimensions shall be provided in connection with any water-engines that may be actuated by the direct pressure from the mains.

(11) In cases where a supply is given by meter, the measuring-apparatus shall be provided and fixed by the Water Authority.

(12) All pipes, taps and other fittings used in connection with a supply of water to any premises shall be similar to samples to be submitted to, tested and approved by, the Water Authority, whose stamp shall be thereupon impressed upon such articles.

398. The foregoing regulations are not to be regarded as either complete in themselves, or as in any sense exhaustive of the numerous conditions which must, in practice, be imposed upon systems of water-supply. They are intended to be merely suggestive of the main lines to be followed in drawing up such a code of rules ; with special reference to a few important practical details, attention to which is absolutely essential to the maintenance of an efficient domestic water-service. General rules are strictly applicable to circumstances between which a close analogy is found to subsist. The complete statement of the best conditions of supply in any given case, can only be arrived at after due consideration of all the features peculiar to it.

INDEX.

LONDON : PRINTED BY WILLIAM CLOWES AND SONS, LIMITED,
STAMFORD STREET AND CHARING CROSS.